Critical
Rationalism

Critical Rationalism

A Restatement and Defence

David Miller

Open  Court

Chicago and La Salle, Illinois

The quotation from page 153 of W.K.C. Guthrie, *Plato: Protagoras and Meno,* Penguin Classics, Harmondsworth 1956, © W.K.C. Guthrie, 1956, is reproduced by permission of Penguin Books Ltd.

OPEN COURT and the above logo are registered in the U.S. Patent and Trademark Office.

© 1994 by D. W. Miller

First printing 1994

Printed and bound in the United States of America.

Library of Congress Cataloging-in-Publication Data

Miller, David (David W.)
    Critical rationalism : a restatement and defence / David Miller.
      p. cm.
    Includes bibliographical references and index.
    ISBN 0-8126-9197-0. — ISBN 0-8126-9198-9 (pbk.)
    1. Science—philosophy. 2. Rationalism. 3. Induction (Logic)
  4. Popper, Karl Raimund, Sir, 1902–   Logik der Forschung.
  I. Title.
Q175.M629 1994
501—dc20   94-11205
                                                       CIP

CONTENTS

PREFACE

The principal thesis of this book is that a rationalist approach to scientific knowledge can be unhesitatingly maintained provided that we surrender completely the doctrine that identifies rationality with justification, whether conclusive or inconclusive. What is rational about scientific activity is not that it provides us with reasons for its conclusions, which it does not, but that it takes seriously the use of reason—deductive logic, that is—in the criticism and appraisal of those conclusions. This is not quite an original thesis, but I hope that I have been able to develop it with thoroughness and consistency, and even to extend it. Although a respect for logic permeates the entire discussion, I have tried hard not to write a technical book. In the early chapters, in particular, I have banished all logical technicalities to short concluding notes. Mathematical considerations break out more persistently only in Chapters 7 and 8, where a little knowledge of elementary probability theory is required; in 10.4, which deals with some logical issues concerned with verisimilitude; and in Chapter 11, which uses some very elementary algebra.

It is an important part of my thesis that since logic is used only to probe, never to prove, all logic is deductive logic. There is no inductive logic. In the first section of *Logik der Forschung* (1934/1959a, 29), Popper wrote:

> Some who believe in inductive logic are anxious to point out, with Reichenbach, that "the principle of induction is unreservedly accepted by the whole of science and that no man can seriously doubt this principle in everyday life either" [here there is a footnote referring to a contribution of Reichenbach's in *Erkenntnis* 1]. . . . Yet even supposing this were the case—for after all, "the whole of science" might err—I

should still contend that a principle of induction is superfluous, and that it must lead to logical inconsistencies.

After briefly describing these inconsistencies, and attributing their discovery to Hume, Popper goes on (29f):

> My own view is that the various difficulties of inductive logic here sketched are insurmountable. So also, I fear, are those inherent in the doctrine, so widely current today, that inductive inference, although not 'strictly valid', *can attain some degree of 'reliability' or of 'probability'*. . . .
>
> The theory to be developed in the following pages stands directly opposed to all attempts to operate with the ideas of inductive logic. It might be described as the theory of *the deductive method of testing*, or as the view that a hypothesis can only be empirically *tested*—and only *after* it has been advanced.

There is not much more in *Logik der Forschung* about the impossibility of constructing a system of inductive logic (though the final chapter, especially Sections 79–81, 83f, contains some devastating criticisms of various proposals of Reichenbach, Keynes, and Kaila). The book is for the most part a superb elaboration of falsificationism—Popper's theory of how science may grow through an involved interplay of dicey conjectures and dusty refutations. With this theory Popper claims to have solved the problem of induction, though as he later acknowledged (1972, 1), "few philosophers would support the thesis" that he has done any such thing. This is sadly no more than the simple truth. Sixty years after the publication of *Logik der Forschung* Popper's daring innovation continues to be widely misrepresented and misunderstood—if not deprecated as a barren form of scepticism, or even irrationalism (for example, by Lakatos 1974, 261; Mellor 1977, 196; O'Hear 1980, 207; Newton-Smith 1981, 51), then depreciated for itself incorporating some principle or rule of induction. In the course of this book I intend to refute all such falsifications of falsificationism. I at least count myself amongst those few who see in Popper's achievement one of the most remarkable, and indeed most exhilarating, developments in the recent history of thought. I hope that this book will be a contribution to a sounder and less blinkered appreciation of Popper's work than it has previously enjoyed.

In Chapter 1 I sketch in an oversimplified way—a determinedly oversimplified way—the achievement of falsificationism; and in the chapter that follows explain why the problem of induction is at last well and truly solved. Chapters 3 and 4 consider more abstractly the logical the logical features of falsificationism and its generalization, critical rationalism, and try to demonstrate that this philosophy is not simply verificationism turned upside down. Chapter 5 is a scherzo, more provocation than real argument, which contrasts falsificationism with a wide range of other proposed solutions to the problem of induction. The three of these that take Hume seriously enough to abandon inductive logic, but not seriously enough to abandon logic, are picked out for more respectful critical attention in Chapter 6. The most popular of them, Bayesianism, which is often represented as a subtle refinement of falsificationism, is then further criticized in Chapters 7 and 8; my business there is to show how methodologically unbusinesslike Bayesianism is, and how far it is from illuminating even the surface features of scientific activity. The remaining chapters of the book are devoted to some more technical issues, probability and verisimilitude, here handled, I hope, without undue technicality. In Chapter 9 I consider a problem that Popper from the start represented as a crucial test case for falsificationism, namely the question of the falsifiability of statements of physical probability; and then go on to discuss the propensity interpretation of probability, which I try to defend from recent criticism and also to challenge. In conclusion, Chapter 10 picks over the ruins of the debate on verisimilitude, and Chapter 11 outlines what I still regard as the single most intractable problem in this area. It is perhaps not an ideal finale to a book whose goal is to defend science, as far as it can be defended, as a rational enterprise. But it is not my wish to suggest that the simple version of falsificationism advocated here is capable on its own of solving all methodological problems. I have no doubt, however, that the problems that still face us, if soluble at all, can be solved in agreement with falsificationist principles.

Some of the material in this book has formed the subject of invited lectures at the following conferences: 'The Applications of Inductive Logic', The Queen's College, Oxford 1978; 'Karl Popper et la Science d'Aujourdhui', Cérisy-la-Salle 1981; the Joint Session of the Mind Association and the Aristotelian Society, Manchester 1981; the Second World Conference on 'Mathematics at the Service of Man', Las Palmas 1982; 'La Filosofía de Karl Popper', Barcelona 1984; 'La realtà ritrovata', Milan 1985; 'Man in the Age of Technology', Athens 1985; Eleventh International Wittgenstein Symposium, Kirchberg am Wechsel 1986; 'Philosophie des Probabilités', Paris 1990; Seventh One-Day Conference on the Philosophy of Karl Popper, London 1990; 'Las Filosofías a Finales del Siglo XX. Crisis y Renovación', Mexico City

1991; and Coloquio de Filosofía de la Lógica, Filosofía del Lenguaje, Filosofía de la Mente y Metafísica, Buenos Aires 1991. I have also drawn a good deal on material presented at various times in talks at the London School of Economics; the Universities of Cambridge, Keele, and Stirling; Middlesex University; the British Society for the Philosophy of Science; CREA, École Polytechnique, and École Normale Supérieure, Paris; Institut d' Estudis Catalans, Barcelona; Istituto di Cibernetica, Arco Felice; the Austrian Economic Association; the Jagiellonian University; the Universities of Helsinki and Turku; the Finnish Philosophical Society; the Society for Critical Scientific Thought, Athens; UAM-Iztapalapa, Mexico City; UNAM, Mexico City; the University of São Paulo [USP]; Universidade Estadual Paulista [UNESP]; Pontifícia Universidade Católica do Rio de Janeiro [PUC-RIO]; and Universidade Federal do Rio de Janeiro [UFRJ]. I am glad to take this opportunity of thanking my hosts in all these institutions for the generous interest that they have shown in my work.

Parts of the book have already appeared in English in *Foundations of Physics, Fundamenta Scientiae, Stochastica* (Barcelona), and *Cahiers du CREA* (Paris), and in books published by the Clarendon Press, Humanities Press, Kluwer Academic Publishers, University Press of America, and Verlag Hölder-Pichler-Tempsky. I thank the publishers for enabling me in this way to reach a wider audience. All previously published work has been revised, some of it extensively, not only to take account of published criticism, but also to impose some degree of uniformity on papers originally written for diverse purposes and diverse audiences. I have not, however, tried to deny the original conversational tone of one of two of the chapters. For the sake of further consistency I have also allowed myself to make inconsequential changes when quoting from the works of other authors.

The intellectual debts of the book are numerous. Special thanks are due to Larry Briskman, with whom many of the ideas in the book were discussed time and time again. It is a pleasure also, at a time when old-fashioned intellectual virtues such as truth, intelligibility, and honesty are treated with so much derision all around, to thank publicly the members of Alain Boyer's group at CREA, including also Daniel Andler, Jacques Dubucs, Bernard Walliser, Denis Zwirn, and Hervé Zwirn. In this part of Paris a genuine enthusiasm for scientific and philosophical problems is kept vibrantly alive. Others whom I need to thank for advice or criticism or encouragement include at least Andrew Barker, Newton da Costa, Joe Eaton, Derek Freeman, Jack Good, Leslie

Graves, Peter Schroeder-Heister, Risto Hilpinen, Wilfrid Hodges, Deryck Horton, Colin Howson, Paul Humphreys, Greg Hunt, Jessica Kingsley, Theo Kuipers, Jan Lester, Isaac Levi, Paul Levinson, Dennis Lindley, Bryan Magee, Patrick Maher, David McDonagh, Hugh Mellor, Karl Milford, Alan Millar, Thomas Mormann, Alberto Mura, Alan Musgrave, Ilkka Niiniluoto, Ray Percival, Michael Redhead, George Rowlands, Tom Settle, Jeremy Shearmur, Jack Sim, Marlene Sim, Jim Smith, Ian Stewart, Settimo Termini, and Raimo Tuomela. It should not need to be said that none is responsible for the use I have made, or neglected to make, of advice that I have been given. I should like to thank also David Ramsay Steele at Open Court, for his advice and, above all, for his patience.

The original forms of Chapters 3 and 4 were written to honour the memories of Grover Maxwell and Bill Bartley, and they continue to pay these tributes in the revised versions published here. Chapters 1 and 2 are based on a paper written to celebrate Karl Popper's 80th birthday, and Chapter 9 is based on a 90th birthday paper. Popper has for so long and to such a degree been the principal source of intellectual encouragement and stimulation in my life that I can only repeat what I said in a dedicatory note appended to that 80th birthday present: "Popper has long shared many of his ideas with me, and I am happy to be able on the present occasion to share some of his ideas with him." If I have succeeded in conveying even some of the excitement that I have experienced and some of the illumination that I have gained in many years of association with this great philosopher, then this book will have been worth writing.

1

Conjectural Knowledge

1.1. The Problem of Induction

The task of empirical science, like that of other investigative disciplines, is to separate as thoroughly and efficiently as it can the true statements about the world from those that are false, and to retain the truths. The first duty of the theory of human knowledge must therefore be to explain how, if at all, such a separation may be achieved. What is not of immediate concern is whether we can know that it has been achieved; that is, whether empirical statements (observational and other) can be established with certainty, or endowed with conclusive and irrevocable empirical verification. The mission is to classify truths, not to certify them.

But it is just for this reason that the pursuit of conclusive verification, if practicable, might after all render a useful, if limited, epistemological service. For any statement verified as true could be classified as true; a statement certified as true is, certainly, true. Unfortunately for such a project, conclusive verification by empirical means turns out, quite trivially, not to be practicable, anyway above the observational level. Universal hypotheses cannot be conclusively verified, even in a finite universe (if only because any claim to have inspected the whole universe is itself a statement of universal form; see Russell 1918/1956, 235; 1918/1985, 101). Thus few, if any, interesting statements will be classified as true if they are required at the same time to be certified as true. Even if it is the case that some statements of observation are successfully self-certifying, it is incontrovertibly the case also that such statements resemble in no respect the hypotheses of theoretical empirical science.

If empirical science is possible, therefore, as it does seem to be, then the segregation of truths and falsehoods is achieved in some other way. A popular idea, hardly limited at all in the variety of its forms, is that we introduce a formal (topic-neutral) principle of induction (or principle of uniformity of nature, or principle of universal causation), itself certified true, that in conjunction with observational statements allows us to certify theoretical statements as true. But no such principle has ever been discovered; not only are the claimants to the title not certifiably or certainly true, they are not even true. Nor, as Hume so poignantly pointed out, is any such principle even possible unless it is itself certified not by empirical means but a priori. Kant suggested just this; but it is seldom thought, even by Kantians, that "his ingenious attempt to provide an *a priori* justification for synthetic statements was successful" (Popper 1934/1959a, 29; see also Popper 1963, 42–48). Similar objections may be raised against the very similar proposals to operate with formal rules of inductive inference that transmit certainty from the observational statements to the theoretical ones. Again, no rules that do this have ever been proposed. Nor are they possible unless their validity is open to some a priori certification. It is not known how any such certification could be achieved.

There will be few to disagree with the conclusion that the prospects in this sort of direction are unrelievedly bleak. For almost everyone these days is a fallibilist; almost no one now supposes that empirical statements, even simple observational ones, can be established with certainty. But as a result of this realization, methodology seems to have taken a wrong turn, reacting as though it was certain truth, rather than truth itself, that had always been the primary goal of scientific inquiry. Instead of looking for new methods of sifting truths from falsehoods, most philosophers have preferred to tinker with the old method: for full verification they have substituted partial verification—that is, confirmation; for full certainty they have substituted partial certainty—that is, probability; for establishment they have substituted support; for conclusive reasons they have substituted merely good reasons (see especially Chapter 3 below). The goal of science has come to be envisaged as the accumulation of highly confirmed, or highly probable, or well-supported hypotheses; and only derivatively as an accumulation of truths. Indeed the old problem, the problem of explaining how we have discovered so much about the world, and how we are to discover more, has been largely neglected, if not abandoned; and, I am afraid, those who have abandoned it have failed to substitute for it any serious methodological issue (see Chapters 5 and 6 below).

I do not pretend that this change of direction, this epistemological refraction, took place on a single historical occasion. But its occurring often does not stop it, however often it has occurred, from having been a terrible mistake. There are two main reasons. The first, which has been relentlessly stressed by Popper (1963, 217–20, 1959a, Appendix *vii), is that it vastly underestimates the pervasiveness of fallibility in science. The fallibility, or corrigibility, of scientific hypotheses is not an incidental irritation, as some philosophers seem to think, negligible to first approximation and to be taken account of only in sophisticated disquisitions. On the contrary, scientific hypotheses (and any methods we may employ in their construction) are unimaginably fallible, unimaginably improbable, unimaginably unlikely to be true; a high degree of confirmation, or of evidential support, or a high probability, is scarcely easier to obtain than is certainty itself. But even if a high probability were manageable for some scientific statements, perhaps some observational statements, that would not on its own help us to say whether those statements were true or false. This is the second main objection (made, for example, in Mott 1980, 182). Although there are, plainly, parallels between probability (that is, partial certainty) and certainty, it is quite wrong to suppose that these parallels allow us to make use of hypotheses that are probably true in the same way that we might make use of hypotheses that are certainly true. Certifying a statement as true is one way (though not the only one) of classifying it as true; in parallel, getting near to certifying it as true is one way of getting near to classifying it as true. But it is not a way of classifying it as true, or of classifying it as nearly true, or anything else like that. (For simplicity, I suppose here that the question of whether a statement is near to the truth, like the question of whether it is true, is an objective one, and an interesting one. But see the critical discussion of verisimilitude in Chapters 10 and 11 below.) And since science proceeds not by getting near to separating true and false statements, but by actually separating them (correctly or incorrectly), the study of methods for grading statements by probability does not at all explain what it is that science actually does. The pursuit of probable truth, unlike the pursuit of certain truth, is indeed quite radically disconnected from the search for truth.

This simple fact seems to be ill understood. Harsanyi, for instance, though acknowledging that *"ideally* the task of science is to come up with true theories", thinks that since "we can never be sure whether any given scientific theory is . . . true . . . , *more realistically,* we may say that the task of science is to identify, among all alternative theories existing in a given field, that particular theory . . . judged to have the *highest probability* of being true" (1985, 11). (See also 5.7.f. below.) But although the search for theories that are highly probable may be marginally more realistic than the search for theories that are certain, it is not in the same league as the search for theories that are true. This applies however probability is interpreted. I agree with Fetzer that recourse to subjective probability in the theory of knowledge "disregards the objective of inquiry" (1981, 221, italics suppressed); but the same holds even if objective probabilities are brought into play.

At this point some philosophers have capitulated to the sceptical attack, and have concluded that if truth cannot be obtained with certainty then it should not be sought at all. Some of them have simply given up altogether, and have gone off to practise mysticism or some other new hobby. Some have turned to relativism and anti-realism, doctrines hardly meriting consideration in the presence of the clear distinction between what is true and how (if at all) we know it to be true. But others, Bayesians as they are now generally called, have tried to understand the scientific enterprise in its entirety with the attenuated apparatus of judgements of probability, with no mention (at least, no mention above the observational level) of truth and falsity. In common with falsificationists, pure Bayesians (there are many impure ones) eschew all attempts to introduce ways of reasoning that are not deductive. For the moment I leave Bayesianism aside, but I shall criticize it in some detail, and I hope destructively, in 6.5, and also in Chapters 7 and 8.

Apart from uncomplainingly accepting them, there are obvious responses to each of the two objections I have raised to the idea that science, unable to attain certainty, must strive for the next best thing: the objection that theories cannot be made interestingly probable; and the objection (which I find much more compelling) that even if they could be, that would do nothing to advance us towards classifying them as true or false. To the first objection it is sometimes said (Jeffrey 1975, 114–16) that it is not the sheer probability of a hypothesis that is of first importance, but the extent to which this probability has been increased by the evidence amassed, or, alternatively, the extent to which it exceeds the probabilities of rival hypotheses. I have doubts whether such adjustments are at all adequate to take account of Popper's insistence on the inverse relation between probability and content; but here I need only note that by so tackling the first of the objections they run themselves more deeply into trouble over the second. Here the natural response is to resort to a principle or rule (sometimes called a rule of acceptance) that lets us move directly from judgements of probability (or of increased probability, or perhaps of superior probability) to judgements of simple truth and falsity. And a variety of such principles and rules can indeed be formulated. But what must be realized of any such principle—call it *P*—is that, unlike its usable analogue in the case of certainty ('All certainly true statements are true'), it is not a logical truth, but at best factual; quite possibly, it is logically false, the fate of the 'rule of high probability' (for which see Hilpinen 1968, Chapter 2). A question that therefore emerges is whether *P* is true; it can hardly be used in the classification of other truths if it is itself false (or of unstated truth value). Now this task clearly is not to be undertaken by first gathering empirical support for *P*, by partially certifying it and then applying *P*; that would immediately initiate an inescapable infinite regress (or else produce a situation similar to the truth-teller variant of the liar paradox). If the principle *P* is to be classified as true, as indeed it must be if it is to be used in further classifications, then we need some way of classifying it that is quite independent of *P*.

The lack of specification of *P* indicates that the above argument is of considerable generality; indeed, it works for any system in which hypotheses are supposed to be able to obtain degrees of probability or empirical support short of certainty. But the lack of specification of *P* is a symptom also of the fact that no one has ever formulated such a principle that anyone else has had the slightest inclination to classify as true. (It is not just that—as fallibilism says—well-supported hypotheses might not be true, but that—as the fate of Newton's theory forcefully shows—sometimes they actually are not true.) Indeed if our goal is simply to sort out what is true, the detour through probability or confirmation, or support, in the company of the imaginary principle *P*, is plainly gratuitous. For rather than spend our time trying to classify some absurdly general principle like *P* as true we would do better to investigate, and try to classify as true (or as false) those much more manageable, yet also much more interesting, factual statements that are at the centre of our concerns, namely genuine scientific hypotheses. Such classification cannot always be done by appeal to empirical support, as we have seen; some classification must precede all certification. Falsificationism proposes that it is never done by appeal to empirical support (so that no principle like *P* is needed). This is not at all to suggest that it is done without reference to empirical evidence, or to the facts. That would be to abandon empirical science. All that is suggested is that we should concern ourselves solely with the truth and falsity of the various statements of science that come to our attention, not with the question of whether these statements have accumulated any empirical support. The central idea of falsificationism is that the purpose of empirical investigation is to classify hypotheses as false, not to assist (with the help of some principle like *P*) in classifying them as true. As I shall attempt to show, recognition of this simple fact leads at once to a most beautiful explanation of how it is that science endeavours to sort out the truths and falsehoods that come within its scope.

1.2. Outline of Falsificationism

Popper is of course a fallibilist. But his central achievement, in my view, has been to recognize that, in contrast to the ideal of truth, the ideal of certainty is quite barren if it is not fully attainable. Science according to Popper, is not, need not be, and should not be, dominated by this ideal of certainty. In rejecting what Dewey called the quest for certainty, Popper is not merely acknowledging the correctness of fallibilism. It is the quest that is being rejected, not the certainty alone. That is to say, the crucial disagreement between Popper and other philosophers is a methodological one, not an epistemological one. We all agree that certain truth cannot be achieved. As far as I know, Popper was the first to say that it ought therefore not to be attempted. As we have seen in the previous section, probable truth, reliable truth, and so on are simply not worth achieving if our aim is truth; nor is there any point in trying to make our knowledge more certain, more probable, more reliable. For if the aim of science is truth alone, then it is on truth alone that the methods of science should be concentrated. It is in this sense that methodology should be minimal, concerned only with truth value. As the appropriate method in the search for truth Popper proposes the method of conjectures and refutations.

The main difference between this method and all justificationist methods—that is, all methods that seek to justify, if only in part, the hypotheses that scientists propose—is that it relies on expulsion procedures, rather than entrance examinations, as the chief way of maintaining academic standards. For justificationists a hypothesis has to pass tests, or be confirmed, or in some other way be touched with grace, if it is to be admitted to the realm of scientific knowledge; if it fails these tests, or is disconfirmed, or even if it fails to be confirmed, it is excluded. Falsificationists, in contrast, think that a hypothesis need submit to tests only after it has been admitted to science. If it fails any of the tests to which it is put, then it is expelled, removed from science; if it passes them all, then nothing happens—that is to say, it is retained. The passing of tests therefore makes not a jot of difference to the status of any hypothesis, though the failing of just one test may make a great deal of difference. For justificationists, on the other hand, the passing of tests is quite as important as the failing of tests, for it is precisely this that determines whether a hypothesis is admitted to the body of science. Of course, justificationists need expulsion procedures as well as entrance examinations; for even the most rigorous entrance examination fails to guarantee the quality of a successful candidate. But the expulsion procedures are the sole means that Popper allows for the control of scientific knowledge. He adopts a policy of open admission, subject naturally to the condition that no hypothesis may be admitted unless there is some way in which, if necessary, it may be expelled. This natural, but essential, qualification is Popper's criterion of demarcation between science and nonscience: a hypothesis may be admitted to the realm of scientific knowledge only if it is falsifiable by experience.

Since our aim is so to shape the body of scientific knowledge that it contains as many truths and as few falsehoods as possible, it is of the greatest importance that the expulsion procedures should be brought into play at every possible opportunity. Open admission would be a foolish policy if it were not complemented by a rigorous campaign of permanent deregistration. Particularly dangerous is the verificationist trickery of emphasizing good points and disregarding bad ones, which is sometimes used to maintain an unworthy but favoured student *in statu pupillari*. (For a critical analysis of the most arresting example of recent years, see Freeman 1983.) Popper's methodology accordingly insists that, if we are seriously searching for the truth, we should submit any hypothesis proposed to the most searching barrage of criticism, in the hope that if it is false it will reveal itself as false. To the extent that we conform to the principle of empiricism, "the principle that only 'experience' can decide about the truth or falsity of a factual statement" (Popper 1933/1959a, 312), our criticism will take the form of empirical tests, of confrontation between the hypothesis under fire and the kinds of phenomena—photographs, behaviour of physical apparatus, and the like—that are generally regarded as making up the empirical basis of science. These empirical tests provide us with a way of classifying as true or false the test statements that truly or falsely report their outcomes. Should a hypothesis clearly and unambiguously fail one of the tests to which it is submitted, should it (that is to say) contradict the test report, then we shall classify it itself as false, and at once eliminate it as a candidate for the truth. For since the aim of science is just the truth, we shall want to retain anything that might be true, but nothing that cannot be true. In this simple situation falsification will lead to rejection, and lack of falsification will mean no change; that is, the hypothesis will be retained. And such hypotheses as do remain after testing, in particular those that remain after crucial tests have eliminated some of their competitors, may indeed be conjectured to be true.

"Science advances conjectures, not certain information", writes one commentator (Ackermann 1976, 3), summing up one aspect of Popper's theory. The statement is of course correct, but it highlights a false contrast. The conjectural character of scientific hypotheses lies not as much in the fact that they cannot be shown to be right as in the fact that they are ready to be shown to be wrong. The falsifiability of scientific hypotheses is indeed of cardinal importance to any methodology that practises a policy of open admission to the domain of science. For unfalsifiable hypotheses, if appointed, would effectively have security of tenure; and such security is, to say the least, undesirable if what we are seeking is the truth. How could we, even in principle, end up with a science full of truths if there are false but unfalsifiable hypotheses entrenched in the woodwork? Thus the falsifiability, even the high falsifiability, of scientific hypotheses is very far from being a feature to be casually dispensed with.

Let us turn for a moment from the theoretical domain, where Popper's solution to the problem of induction has often been accepted by inductivists, to the practical domain, where Popper's solution to the problem of induction has been as often rejected by anti-inductivists. Many have maintained that, however defensible Popper's deductivist theory of testing may be in the realm of theories, in matters of practice and practical action a principle of induction becomes essential; to act rationally we cannot do other than put our trust in hypotheses that are reliable. Popper's advice on this matter has been given as follows (Popper 1972, 21f).

> From a rational point of view we should not 'rely' on any theory, for no theory has been shown to be true, or can be shown to be true.

> . . . But we should *prefer* as basis for action the best-tested theory.
> . . . the best-tested theory is the one which, in the light of our *critical discussion*, appears to be the best so far . . .
> . . . in spite of the 'rationality' of choosing the best-tested theory as a basis of action, this choice is *not* 'rational' in the sense that it is based upon *good reasons* for expecting that it will in practice be a successful choice: *there can be no good reasons in this sense, and this is precisely Hume's result.*

In other words, in the lucky event that there is an unrefuted hypothesis relevant to our practical problem, we shall normally plan our actions on the assumption that this hypothesis is true. Nothing, of course, justifies our action, let alone our theory, but this does not make our action wrong or irrational. As I say, these proposals of Popper's have been almost universally condemned as involving some covert principle of induction. In item 2.2.g, and again in 3.4, I shall show that they need involve nothing of the kind. By thereby repudiating what is generally regarded as the most trenchant of all criticisms of falsificationism, I hope to render visible its hidden strengths.

In brief, Popper offers us the following refreshingly uncluttered picture of how science proceeds. In order to discover something true we propose conjectures, which might be true, and incorporate them into science, without regard for whether there is anything that could be called evidence in their favour, without regard for whether there is any reason to think that they are true. We then make the most ruthless and uncompromising efforts to show that these hypotheses are not true, and to reject them from science. It is of the utmost importance that we should not be hampered in our attempts to do this, either by the form of the hypotheses themselves or by the deployment of artifices raised in their defence. Not only must our hypotheses be open to falsification by experience, our treatment of them must be determinedly one of eliminating them if in fact they are false. For although error is not something to be feared if one is concerned to attain the truth, irrevocable error certainly is; it is not the perpetration of error, but its perpetuation, that we must adamantly oppose. (This holds in practice as well as in theory. Thus a revolutionary approach to theoretical science and a piecemeal approach to its application are in harmony and not, as some may suppose, at loggerheads.) Hypotheses that pass all the tests to which they are subjected are retained, and there is no call to reconsider their conjectural classification among the truths. Hypotheses that fail tests are classified as false. Since we are interested in as obtaining as much truth as we can, we shall prefer, among the unrefuted hypotheses, the logically stronger or more informative ones, which subsume those that are logically weaker (and, if false, are more easily shown to be false). There is no need to worry at any stage whether the retained hypotheses have any evidence in their favour, but there is always a need to worry whether we have subjected them sufficiently to tests.

This simple picture may be complicated in a number of ways, but the principles remain the same. One may note, for example, as did Popper (1960/1963, 231–36; 1972, 57f), that science often operates as though it aims not at truth but merely at close approximations to the truth. In such circumstances a hypothesis will not be discarded as soon as it is falsified, but only when it has been shown to approximate the truth worse than does some rival hypothesis. A related complication, not easily resolved, stems from the fact that rival hypotheses are rarely logically comparable; hence judgements of relative informative power often have to be approximate too. On the requirement that statements of relative approximation to truth be falsifiable, see Chapter 10 below; on the difficulty of comparing theories by content, and even numerical theories by accuracy, see also Chapter 11. One may also note, as Popper began to note rather early on (1940/1963, especially 313, and 1945a/ 1957, 121), that hypotheses are admitted to science not in an utterly blind manner, but only if they make some attempt, however feeble, to solve some pressing scientific problem. Again, failure to provide any such solution will be deemed reason for dismissal, even in the absence of empirical falsification. (It should be recognized that although the demand that a hypothesis be directed at some problem functions rather like an entrance examination, it is decidedly not a test of academic merit, but only one of motivation. It is the unserious applicant who is excluded in advance, not the ungifted one.) Or one may note, as do Agassi (1975), Watkins (1975), and others (including Popper himself in Section 33 of his 1974a, and in Chapter IV, 'A Metaphysical Epilogue', of his 1982b), the pervasiveness of metaphysical and other untestable hypotheses within science. Such hypotheses are often introduced as essential adjuncts to scientific hypotheses; indeed, all falsifiable hypotheses have amongst their consequences a host of unfalsifiable statements (ranging from tautologies and unrestricted existential statements to meaty metaphysics) that enter science as it were on the coat-tails of their parents. But these unfalsifiable consequences—to the extent that that is all that they are—are not scientific in their own right; their title is one of courtesy. If their parents are rejected from the realm of scientific knowledge, they will have to be rejected too. Thus the metaphysical component of science need not be denied. It simply needs properly and responsibly to be taken care of. (For a generalization of this point, see 4.3.d below.)

Finally, one may emphasize, as Popper did from the very beginning, that an experimental falsification is not just a matter of confronting a hypothesis with refractory evidence; more often than not it involves the subtle and delicate task of finding out what is actually happening in the experiment. Indeed, no falsification is conclusive, if only because all test statements are themselves fallible and open to dispute. Kuhn's famous claim that "[no] process yet disclosed by the historical study of scientific development at all resembles the methodological stereotype of falsification by direct comparison with nature" (1962, 77) does not affect the logical situation, even if it is true (on which point see Popper 1983, xxv–xxx). But it would be wrong to infer from this that no hypothesis can be properly falsified (Achinstein 1968, 168), and therefore that any hypothesis may be admitted into science and permanently retained (Lehrer 1980, 131f). On the contrary, that a falsification has not been done conclusively does not mean that it has not been done properly. Moreover, if we are not prepared to eliminate hypotheses from science (for fear of making a mistake), then we must not admit any hypotheses to science either. That is the burden of the criterion of demarcation. It is better to have no hypotheses at all than not be able to control them. Indeed science must not be supposed to be under an obligation to hypothesize, or to issue predictions. There is nothing meritoriously scientific about singular predictions that are detached from any theoretical structure (following Popper 1945a/1957a, Section 28; 1963, Chapter 16, they might well be called prophecies). Singular predictions of observable occurrences are testable of course, but often too late to tell us anything of any service. To take an extreme example: the prediction of the result of a lottery is a testable conjecture, but testable only when the draw is made; it is in no interesting sense a scientific hypothesis.

We see therefore that science is a collection of statements, and that the business of science is the discovery, as far as is practicable, of the truth values (and perhaps of the relative degrees of approximation to the truth) of these statements. The whole business can be explained, quite satisfactorily, without any reference to certainty, probability, confirmation, support, reliability, confidence, justification, good reasons, or knowledge. Truth and falsehood suffice. How is it then that Williams, in reconstructing what he calls the project of pure inquiry, "namely that of *trying to find the truth*", can almost effortlessly conclude that "the pursuit of certainty is the only possible road for the *pure* search for truth" (1978, 34, 49)? The answer is that for Williams (and, he suggests, also for Descartes) the search for truth is the search for true beliefs, not true statements. At a crucial point he writes (*op. cit.*, 39):

> Now if I want to acquire a collection of flints, and only prehistoric flints, one way is to collect a lot of flints, and then investigate which of them, if any, are prehistoric. This might be inefficient, compared with a method of acquiring them in the first place which made it more likely that any flints I acquired were prehistoric ones. But the analogous process with acquiring true beliefs would be not just inefficient, but incomprehensible. Since to believe something is to believe that it is true, to acquire a belief is already to assume an answer to the question of whether it is true. So a method which *A* uses as an enquirer . . . must be a method of acquiring beliefs which itself makes it likely that the beliefs *A* acquires by it will be true ones.

What Williams is saying in effect is that the process of falsification and error elimination cannot be carried out wholly within the realm of beliefs or convictions. This is a point of the greatest significance (though Williams himself draws absolutely the wrong conclusion from it). It shows to what extent scientific activity, if it is to be the pursuit of truth, must be conducted in the objective—but humanly created—world of statements-in-themselves (Popper 1972/1979, especially Chapters 3 and 4). This objectivity is important if we are not to be misled by the truth of fallibilism into thinking that errors are always with us and effectively beyond our control. Incidentally, it is a remarkable thing that Popper's doctrine of the objectivity of scientific knowledge and scientific method can be understood, as it is by Haack (1979), as an attempt to eliminate human beings altogether from the development of scientific knowledge. Haack judges the attempt to be unsuccessful (as well she might) on the grounds that genuine fallibilism must be concerned with beliefs—it does not make sense if restricted to an objective domain of statements; to which it is a sufficient response that methods, which are objective, can be fallible too. In any event, neither trousers nor trouser repairs are in the mind, but it would be wrong to infer that tailoring is not a human activity.

1.3. Conclusion

Hume showed that inductive arguments could not be justified, even in part, but he did not think that they were thereby incorrect. Most later writers have agreed. It seems, therefore, that if we could only bring ourselves to abandon justificationism, the ever-vulnerable doctrine that rational men hold only to doctrines that they can justify, then induction could be rescued. (This is the viewpoint of Jones and Perry [1982], who maintain that if falsificationism needs no justification then inductivism needs no justification either.) Following the lead of Popper, and also of Bartley (1964; 1984), I have urged above that justificationism can be given up without lapsing into scepticism or irrationalism, a topic vigorously pursued in Chapters 2 and 3. (The special features of Bartley's position are considered at length in Chapter 4.) But it is untrue that this sacrifice (I know that most will see it as a sacrifice) permits induction to be saved. For it has turned out to be hopelessly unclear what inductive arguments actually are, even to those who believe in their existence. When Hume suggested that all reasonings from experience would have to "proceed upon that principle, *that instances, of which we have had no experience, must resemble those, of which we have had experience, and that the course of nature continues always uniformly the same*" (1738, Book I, Part III, Section vi, 89), he cannot have seen how flexible that principle was, and that it could offer sanctuary to almost any generalizing argument, however absurd its conclusion. There are only some respects in which "nature continues always uniformly the same". Which they are is exactly what science tries to find out, by the formulation and testing of universal laws. For those who are not justificationists, what Goodman calls the new riddle of induction (1955, Chapter III), the problem of saying which inductive inferences are valid and which are not, is no more and no less than the problem of empirical science: to discover the truths, in particular the universal truths, of this world.

There is of course a sense in which we generalize from experience. But experience has to be reported first, in terms that are already limbering up for generalization. We can only guess at which generalizations will in fact be successful. Call this induction if you must, as does Good (1975, 61; 1983, 163f), but do not derange the discussion further by pretending that at its heart there lies anything that can be called an inductive method or an inductive logic.

2

Popper's Solution of the Problem of Induction

2.1. Enumeration of Objections

Falsificationism has often been objected to on the grounds that, contrary to its intentions, it needs to adopt some principle or rule of induction if it is to succeed in explaining how scientific knowledge grows. These objections, all of which will be answered in this chapter, may be loosely collected into nine categories. I shall enumerate them first and eliminate them second. Categories *a.–d.* are assuredly the more fundamental; but as already noted, it is on *g.*, the arena of practice, that the majority of anti-falsificationist hopes are now pinned.

a. *The Presuppositions of Science*

The first, and perhaps most fundamental, objection is found in the claim that science, and even pre-scientific inquiry, cannot get started without some assumption or presupposition of order and regularity in the world. This objection was considered, and set aside, already in *Logik der Forschung*, Section 79, but it recurs from time to time. O'Hear, for example, writes (1980, 57f and 60f):

> Popper's attempt to dispense with induction is unsuccessful. We have found that inductive reasoning, removed from one part of the picture, crops up in another. . . . the underlying reason for this is that any coherent conceptualization of experience requires the assumption of a stable order in the world. . . .
>
> The rationality of induction . . . is of a piece with the rationality of belief in an external, objective world.

Feyerabend perhaps voices something like the same objection (1978, 161f):

> Nor can the difficulties of . . . inductivism be overcome by the method of conjectures and refutations. . . . There is not a single method that

> works in all circumstances . . . and that can start without much prepara-
> tion. The method of conjectures and refutations, for example, gives
> results only in a world that is not immersed in arbitrarily distributed
> disturbances (for otherwise a law would be refuted the moment it is
> formulated and a science could never arise). . . . Now once one admits
> that the objections against inductivism can be raised against every
> methodology then philosophy or logic alone cannot decide the matter.
> The matter can only be decided by resolutely using the method and
> seeing where it leads.

In the same vein Trusted tells us that "the very notion of an expectation
of order points to an inductive generalising from experience" (1979,
63).

b. The Content of Science

Next most fundamental are those objections that assume that all our
knowledge of the world must be either direct (observational) or
inferred; more particularly, if we know anything of the unobserved then
what we know must have been obtained by means of some process of
inference. Salmon, for example, writes (1968, 26–28):

> If science is to amount to more than a mere collection of statements
> describing our observations and various reformulations thereof, it must
> embody some other methods besides observation and deduction.
> Popper has supplied the additional factor: corroboration. . . .
>   Corroboration is, I think, a nondemonstrative kind of inference. It is
> a way for providing for the acceptance of hypotheses even though the
> content of these hypotheses goes far beyond that of basic statements.
> Modus tollens without corroboration is empty; modus tollens with
> corroboration is induction.

Salmon has repeated this criticism (1978, 11f), but now seems to be
ready to withdraw it (1981, 119). Similar objections have been made by
Good (1975, 61; 1983, 163f) and O'Hear (1975; 1980, 45f), who
suggests that any use of universal hypotheses is covertly inductive.

c. The Empirical Basis

It has become fashionable to claim that rules of induction are needed
even for the classification of test statements as true or false, that the
acceptance of basic statements is unavoidably an inductive step. Hübner,
for example, writes (1978, 280):

But every falsification, too, we can reply, has some premises, such as axioms of certain observational theories. Now if these premises are conjectural . . . the falsification is conjectural too. This conjecture may be purely arbitrary, and consequently the falsification would be practically meaningless; or the scientist has some reasons for his conjectures; but in that case he cannot avoid using inductions either.

Newton-Smith asks similarly what are "the ultimate grounds on which hypotheses are shown to be false within the Popperian model of science" and, concluding correctly that there are none, goes on to assert that "there can be no grounded falsification without induction" and that "rejection of theories for Popper . . . [is] a matter of mob psychology" (1980, 152 and 1981, 64). More recently, Watkins has maintained that "a solution to [the] . . . problem . . . of show[ing] that there may be rational acceptance of level-1 [basic] statements . . . has not been provided by Popper" (1984, 254), and has attempted to provide a non-inductivist solution himself.

d. The Repeatability of Tests

The fourth species of objection is concerned with the repetition and repeatability of tests. A classic objection in this genre is that of Ayer (1956, 74):

But even if . . . [falsificationism] is the correct account of scientific method it does not eliminate the problem of induction. For what would be the point of testing a hypothesis except to confirm it? Why should a hypothesis which has failed the test be discarded unless this shows it to be unreliable; that is, except on the assumption that having failed once it is likely to fail again? It is true that there would be a contradiction in holding both that a hypothesis had been falsified and that it was universally valid; but there would be no contradiction in holding that a hypothesis which had been falsified was the more likely to hold good in future cases. Falsification might be regarded as a sort of infantile disease which even the healthiest hypotheses could be depended on to catch. Once they had had it there would be a smaller chance of their catching it again.

Ayer clearly thought that this rhetorical question "What would be the point of testing a hypothesis except to confirm it?" was unanswerable by falsificationism, and (if asked often enough) sufficed to refute it, for he repeated the question in nearly the same words on several later occasions

(1972, 74; 1973, 158; 1982, 134). In the same vein Hesse writes (1974, 95):

> Again, one past falsification of a generalization does not imply that the generalization is false in *future* instances. To assume that it will be falsified in similar circumstances is to make an inductive assumption, and without this assumption there is no reason why we should not continue to rely upon all falsified generalizations.

Very much the same point has been made by Warnock (1960), by Levison (1974, 328–330), by Trusted (1979, 63f) and by O'Hear (1980, 63). Schlesinger goes even further (1988, 168f) when he writes that:

> the scientist does not ever take a single step without relying on induction. Indeed, his unbridled commitment to the principle that the future will be like the past, that is, the principle of the uniformity of nature or briefly the principle of induction, is so all-embracing that even when he wishes to falsify a hypothesis he does so on the basis of his firm conviction of the validity of that principle. . . .
>
> Clearly, therefore, a scientist could never believe [himself] to have falsified any theory unless he made the *entirely unprovable* assumption that whatever has been observed to be false here and now is false everywhere, at all times; that the world can be relied upon not to change from place to place.

e. The Law of Diminishing Returns

The fifth variety of objection is also concerned with the repeatability of tests, but it focuses specifically on their severity. As stressed above, Popper requires that the hypotheses we propose be submitted to the most strenuous tests we can devise, in the hope that any hypothesis that is false will be falsified. Very roughly, a test is severe if one of its possible outcomes is very much to be expected in the light of the hypothesis under test but very much unexpected in the light of background knowledge alone: the hypothesis differs dramatically, as it were, from common sense. Popper has written (1963, 390f):

> A serious empirical test always consists in the attempt to find a refutation, a counter example. In the search for a counter example we have to use our background knowledge; for we always try to refute first the *most risky* predictions. . . . Now if a theory stands up to many such tests, then, owing to the incorporation of the results of our tests into background knowledge, there may be, after a time, no places left where

(in the light of our new background knowledge) counter examples can . . . be expected to occur. But this means that the degree of severity of our test declines. This is also the reason why an often repeated test will no longer be considered as significant or severe: there is something like a law of diminishing returns from repeated tests.

Contesting this view of severe tests, Hesse writes (*loc. cit.*):

> Objections can be made to Popper's view on the grounds that it is impossible even to state without making some inductive assumptions. For example, it is not clear that the notion of a 'severe test' is free of such assumptions. Does this mean 'tests of the same kind that have toppled many generalizations in the past', which are therefore likely to find out the weak spots of this generalization in the future? Or does it mean 'tests which we should expect on the basis of past experience to refute this particular generalization'? In either case there is certainly an appeal to induction.

Contesting the claim that there is a law of diminishing returns that does not depend on induction, O'Hear writes (*op. cit.*, 45):

> From an anti-inductivist standpoint, however, the fact that a theory has survived a certain type of test on occasions can give us no reason to suppose that it is more likely to survive another test of the same type on the $n+1$ occasion than it was on the first occasion of undergoing a test of that type. . . .
> In determining test severity background knowledge does appear on any reading to be used inductively.

Pretty well the same objection has been voiced by Musgrave (1975, 250f). Musgrave's criticism is endorsed by Grünbaum (1976a, 236f), and by Watkins (1978, 356–360), though neither appears to accept Musgrave's solution. Nor do I, as I shall explain in the next section.

f. Goodman's Paradox

The sixth sort of objection that I have in mind is that that claims that Popper's theory, no less than any inductivist theory, is troubled by problems raised by Goodman's paradox (presented on pages 74f of his 1955 in terms of the up-to-that-time unfamiliar predicate 'grue'). Vincent, for example, summarizing a discussion of Popper's theory of how one statement can corroborate (or confirm) another in the presence of a third, writes (1962, 162f):

> From the foregoing we may conclude that according to Popper's [theory] . . . '*a* is green' confirms 'All emeralds are grue' relative to '*a* is

an emerald and *a* is examined before time *t*.' . . . Indeed, we may
conclude that '*a* is green' confirms the conjunction of 'All emeralds
examined before time *t* are green' and any other logically contingent
hypothesis relative to '*a* is an emerald and *a* is examined before time *t*.' I
take it as obvious that these results are unsatisfactory.

In other words, according to Popper's theory many conflicting hypotheses may be equally well confirmed by exactly the same evidence. Kyburg
delivers the judgment that "Vincent has shown that Popper's criteria,
like everyone else's, founder on Goodman's reef" (1970, 158). As
Bartley (1981, 347f) has pointed out, many writers appear to agree with
this assessment. A recent example is Worrall 1989b; Section 2 of his
paper can be summed up in the single rhetorical question "What *reason*
can a Popperian give for barring the gruesome hypothesis in advance of
any further evidence?" (273).

This criticism, to be sure, contains no direct accusation that Popper's
theory of testing and corroboration is inductivist despite its intentions. Certainly the criticism is not that falsificationism is compelled
after all to depend on a principle of uniformity of nature. (That
would be to confuse the symptoms of the affliction with its cure.)
But the accusation is there in the form: there can be no solution to
Goodman's problem, to Goodman's new problem of induction,
that does not distinguish between soundly and unsoundly made
predictions—that is, between sound and unsound inductive inferences.

g. The Pragmatic Problem of Induction

The seventh department in which critics have been prone to see the need
for a principle of induction has already been mentioned: the context of
practice. The criticism is very old; almost as old, Feigl (1974/1981,
14f) says, as falsificationism itself:

> Karl Popper, deeply impressed with Hume's arguments, abandoned all
> efforts toward a justification of induction; he even denied the impor-
> tance, if not the very occurrence, of induction in the growth of
> knowledge. But, as perhaps the first to criticize this view of Popper's, I
> asked the crucial question why we should put our trust in (or 'place our
> bets on') laws, hypotheses and theories which, despite severe tests, have
> thus far not been refuted. To this question Popper has never given a
> satisfactory answer. . . . he does not provide any reason whatever for the
> generally accepted practice of using a well-corroborated theory as a

guide for further research or, in its practical applications, for our expectations and actions.

In one form or another this criticism is still with us. Cohen, for example, has written (1978a):

> If the aim of scientific enquiry is to give us power over Nature, we cannot do without some way of appraising the evidential justification for relying on a given scientific hypothesis. If we go up in a new kind of plane, or take a new kind of medicine, we want there to be adequate test-results to show that it is reasonably safe.

And then a little later (1978b):

> Popperians of course accuse their opponents of misunderstanding and confusion. But the dilemma here is such a simple one that there is really no room for any misapprehension. Either Popperian science is consistently anti-inductivist, thereby cutting itself off from technological reasoning. Or it adopts a criterion of evidential support, thereby sacrificing its anti-inductivism. . . .
> *Logik der Forschung* is impaled on the first horn of this dilemma.

And once more (1980a, 492):

> But though Popper has thus gone further down the Humean road than anyone else, his philosophy . . . has notoriously been a recipe for a quite unusable kind of pure science. It has been unable to depict a plausible form of rationality that can be attributed to *technology,* i.e., to the selection of . . . [those] scientific hypotheses that are to be relied on for our predictions about the behaviour of physical forces or substances in particular cases. You would be rather rash to be the first person to fly in a plane that had been made in accordance with the boldest conjectures to survive ground tests on its materials, if extreme conditions covered by the bold conjectures were not present in any of the tests. The conjectures would have relatively high Popperian corroboration but be ill-supported by the evidence. The reasons for accepting them would be rather poor.

"The question", according to Salmon (1981, 121), "is whether the scientific approach provides a more rational basis for prediction, for purposes of practical action, than do . . . other methods" (he mentions astrology and numerology as alternatives). Niiniluoto and Tuomela sum up the matter (1973, 203):

> The fundamental problem which Popper fails to answer is this: why is it *rational* to base one's practical decision upon the best-tested theory, if there are *no good reasons* for expecting that it will be a successful choice?

Similar criticisms are voiced by Lakatos (1968, 390–405); Howson and Worrall (1974, 368); Putnam (1974/1981, Section 2); Jeffrey (1975, 111); O'Hear (1980, 36–42); and by many others.

The line of argument is, apparently, almost as alluring to those who dislike induction as it is to those who like it. Watkins, for example, opens an involved discussion of his own non-inductivist solution to the pragmatic problem of induction by insisting on the need for reliability, and he then sketches the position that he had previously held (1984, 337f):

> What an agent would wish the hypotheses that guide his actions to have is not such properties as depth and unity, which the theoretician values, but *reliability*. . . . But . . . what reason, if any, is there for an agent to act on best (or well) corroborated theories and hypotheses? . . . corroboration appraisals tell us nothing about future performance . . . [but, other things being equal] . . . the decision maker may as well fall back on the best corroborated hypothesis since he has nothing else to go on. Popper has suggested that there is nothing *better* to go on than the results of science.

Proceeding to cite approvingly Salmon's remarks, just quoted, and Salmon's claim that for "the Humean Skeptic . . . none of these methods can be shown either more or less rational than any of the others" (*loc. cit.*), Watkins almost immediately concedes defeat, declaring "Game, set and match to Salmon!" (341). A Pyrrhonic victory indeed. Another author impressed by Salmon's overt justificationism and inductivism is Worrall, who asks plaintively of those who think that Galileo's law advises against springing off tall buildings (1989b, 265):

> as . . . [Miller] and Popper have often emphasized, *all* the universal theories we have so far articulated . . . may well turn out to be false. . . . we have no reason to think that Galileo's law won't be falsified in exactly the *next* instance. So why exactly doesn't the advice . . . [given]— namely to act as though it *won't* be the case that Galileo's law will be falsified in the next instance . . . —why exactly doesn't that advice 'involve recourse to a principle of induction'?

It is not quite clear (since his ideas are presented in the form of a dialogue), but Worrall's conclusion (271f) appears to be that the rationalist must, for want of anything better to do, assert dogmatically

> the principle that 'it is rational to take the best corroborated generalisations . . . as . . . guide to future action.' . . . if you *are* going

to bridge the acknowledged deductive gap between past and future . . . then you must make some substantive inductive principle part of your theory of rationality.

All this seems to me to be quite wrong, to be playing into the hands of the sceptic and the irrationalist. In the next section I shall show how naturally and easily falsificationism explains how rational action is possible without the smallest concession to inductivist magic.

b. Methodology and Verisimilitude

The last two types of criticism to be reported here concern science as a search for verisimilitude rather than simply for truth. The first was resorted to by Lakatos in an attempt to bolster his doctrine, mentioned above, that in practical affairs falsificationism is irremediably contaminated by induction. (A brief summary of Lakatos's position appears in Zahar 1982/1983, 167.) The criticism was that, even in purely theoretical concerns, Popper's methodology was in need of some inductive principle. Lakatos (1974/1978, 156–59) wrote that Popper's theory of verisimilitude makes it

> possible, for the first time, to define *progress* even for a sequence of false theories. . . . But this is not enough: we have to *recognise* progress. This can be done easily by an inductive principle . . . which reinterprets the rules of the 'scientific game' as a—conjectural—theory about the . . . *signs of growing verisimilitude of our scientific theories.* . . .
>
> Popper's methodological appraisals are interesting primarily because of the hidden *inductive assumption* that, if one lives up to them, one has a better chance to get nearer to the Truth than otherwise. . . .
>
> Popper . . . still will not say unequivocally that the positive appraisals in his scientific game may be seen as a—conjectural—sign of the growth of conjectural knowledge; that corroboration is a *synthetic*— albeit conjectural—measure of verisimilitude.

In an accompanying footnote (*op. cit.*, 165) Lakatos writes of this "inductive principle" that he prefers it

> in the form that—roughly speaking—the methodology of scientific research programmes is better suited for approximating the truth in our actual universe than any other methodology.

He conceded that such a principle is "sadly irrefutable". It is comforting, therefore, that neither it nor anything resembling it is needed in the methodology of science.

i. The Miracle Argument

The final objection to be considered is a slightly technical one, and not at all easy to get clear. It is prompted by some remarks of Popper's to the effect that "it would be a highly improbable coincidence if a theory like Einstein's could correctly predict very precise measurements not predicted by its predecessors unless there is 'some truth' in it" (1974b, 1192, Note 165b). Popper notes that the fact that a hypothesis makes many true predictions can hardly be taken as showing that it is true (as apparently it was by Whewell [1847, 63] and by Poincaré [1903/1952, 95; 1913/1963, 100f]); for indeed, Newton's theory and Einstein's share very many true predictions yet contradict each other, so they cannot both be probably true. He suggests (1972, 102), however, that

> good agreement with the improbable observed result is neither an accident nor due to the truth of the theory, but simply due to its *truthlikeness.*
>
> This argument . . . would explain why many incompatible theories can agree in many fine points in which it would be intuitively highly improbable that they agree by sheer accident.

There is a similar statement on page 263 of Popper 1981. Truthlikeness is here meant only comparatively: good observational agreement indicates only that the theory in question is closer to the truth than its predecessors are. Popper stresses that the argument is "typically non-inductive", but then concedes (1974b, 1192, Note 165b) that

> there may be a 'whiff' of inductivism here. It enters with the vague realist assumption that reality, though unknown, is in some respects similar to what science tells us or, in other words, with the assumption that science can progress towards greater verisimilitude.

Despite Popper's caution, this series of remarks (which has much in common with what has been advanced in recent years under the flag of scientific realism) has generally been taken to commit him to the nonsense of inductivism that his philosophy is designed to avoid. O'Hear, for example, writes (1980, 67):

> It is not surprising that some commentators have seen this passage as an enormous concession by Popper to his critics. Certainly, with its "whiff"

of inductivism, it brings Popper's philosophy more in line with common sense and with what one would want to say about our knowledge of the world, but it also means that he can no longer complain about corroboration and theory testing being taken by his less cautious readers to have inductive implications.

And Newton-Smith asserts that "it is just false to say that there is a whiff of inductivism here—there is a full-blown storm" (1981, 68). As we shall see, this is just false.

This concludes my exposition of nine ways in which Popper's methodological proposals have been thought to be tinged with inductivism. I shall now show how unforcedly but forcefully falsificationism can parry each one of these nine accusations.

2.2. Elimination of Objections

a. *The Presuppositions of Science*

The first objection, that "scientific method presupposes the *immutability of natural processes*", was, as I have noted, already formulated and set aside in *Logik der Forschung*. Popper argued there that the principle in question is a metaphysical one, best replaced by the "methodological rule [that] . . . natural laws . . . are to be invariant with respect to space and time" (1934/1959a, 253); a rule of method that, he maintained, is easily distinguished from a rule of inference like the rule of induction. It seems to me that this answer, correct as far as it goes, can be expanded. The crux of the matter is that, in order to provide genuinely interesting knowledge of the world, inductivism needs to assume that there is some regularity in the world, whilst falsificationism requires only that there is some regularity in the world—but it need make no assumption to this effect.

In fact, inductivism has to assume quite a bit more—namely that all or most apparent regularities are genuine regularities—and as Popper observed, this has to be treated as unfalsifiable, despite its being pretty obviously false. But all that is needed for falsificationism to be successful (apart from luck, ingenuity, and so on) is something much weaker; something like the truth of 'Every natural event can be subsumed under some natural law'. (Even this is far too strong for indeterminists.) Such a Kantian principle, which, I stress, is not assumed or presupposed by falsificationism, is clearly unfalsifiable and metaphysical (in the terminology of Watkins 1958, it is an all-and-some statement); and actually a restricted version of it may be regarded as a consequence, or near-consequence, of our accumulated scientific knowledge. As was pointed out towards the end of 1.2, our scientific theories invariably have consequences that are unfalsifiable and metaphysical.

Now although falsificationism does not make any metaphysical assumption concerning the immutability of natural processes, Popper does recommend the adoption of the corresponding methodological rule: to search for spatio-temporally invariant laws. And one might suspect that, were the metaphysical principle false, the methodological rule might easily lead us into irrevocable error. But a moment's reflection should show that this is not so: much as we may try to subsume all natural events under laws, we may persistently fail. Thus the rule that Popper proposes will not lead us to classify as false a hypothesis that without it we would classify as true, nor to classify as true a hypothesis that we would, in the absence of the rule, classify as false. To be sure, it may lead us to propose hypotheses that we might otherwise not have thought of; I myself cannot see this as a fault, or as in any way hampering the pursuit of truth.

But not only is it possible that some events are not governed by natural laws, it is perhaps conceivable that lawlessness reigns supreme instead, as Feyerabend suggests (see Sudbury 1973; 1976). Here we see another advantage of falsificationism over inductivism. For although in the absence of regularity the method of conjectures and refutations may yield meagre fruit (apart, perhaps, from a conjecture concerning the lack of regularity), inductivism will be even worse off, being left high and dry with a false (and unfalsifiable) assumption; or, in the case of an approach such as Reichenbach's (Salmon 1967, 52–54), with the ineliminable debris of a transcendental argument gone apocalyptically wrong (Popper 1983, 339). Certainly it is possible that the method of bold conjectures and severe tests could fail to produce much true theoretical knowledge; as Popper himself says, "there are many worlds, possible and actual worlds, in which a search for knowledge and for regularities would fail" (1972, 23). But tampering with the method of conjectures and refutations, by limiting the conjectures or tempering the tests, is not going to make the acquisition of knowledge in such worlds any easier. On the contrary, a speculative and critical approach may encourage us to try out specific conjectures that there is no knowledge to be found; just as the search for, and continued failure to discover, universal laws in the realm of chance phenomena has helped us to see that probabilistic laws are the best that we can expect in certain domains. Feyerabend's own suggestion at this point—that the value of a method "can only be decided by resolutely using the method and seeing where it leads"— sounds to me uneasily like the method of conjectures and refutations; though if logic is abandoned, as Feyerabend hints it might have to be, a vision of where a method leads (if, indeed, it can be supposed to lead anywhere) will hardly enable us to decide what its value is. Indeed, it is not too much to say that here Feyerabend "sounds like a thief who chides his victims for lacking the items he has just taken from them" (Feyerabend 1981, 59). By all means resolutely use a method and see where it leads; but seeing where it leads, and paying attention to what is seen, is simply the method of conjectures and refutations.

What I call above the crux of the matter—that inductivism has to assume regularity, while falsificationism only requires it to be in place—has puzzled some readers, notably Worrall, who confesses that he cannot for the life of him "see what credit accrues from abstaining from an assumption, when one's theory requires the world to be such as to make the assumption true" (1989b, Note 19, 294). I hope that it is now clear that falsificationism incorporates no methodological rule (or 'theory') that requires any assumption of cosmic regularity to be true. To be sure, some assumption of regularity might be necessary if we wanted to predict that the method of conjectures and refutations will be successful. But falsificationists attempt no such prediction (Popper 1972, 23); not least because the assumption, even if necessary, would be woefully insufficient. Falsificationists do not assume either, as part of methodology, that the atmosphere contains oxygen, even though, according to current theories about intelligent life (which may be wrong), the success of falsificationism as a method "requires the world to be such as to make the assumption true".

In sum, science need contain no metaphysical assumption concerning the immutability or order of nature. It need contain no assumption not explicitly available for testing (though it will inevitably have consequences that cannot be tested). Scientific hypotheses propose order for the world; they do not presuppose it.

b. *The Content of Science*

The second objection, to which Salmon, Good, and O'Hear were instanced as parties, was that without some process of inductive inference we can have no knowledge of the world beyond the content of our own observations. To this the falsificationist reply is unaffectedly simple: most of our knowledge, particularly our theoretical knowledge, is obtained neither by observation nor by inference, but by guesswork. You may of course, like Stove (1982), prefer not to dignify as knowledge what is obtained by guesswork, but that decision has the uncomfortable consequence that most scientific knowledge is not knowledge. (See 3.1 below.) On the other side, you may opt to call guesswork a process of inference, but if you do this you must bear in mind that such inferences are not performed according to any rules, and that they are not in need of justification. You may even refer to guesses of universal form as the conclusions of inductive inferences, despite their not proceeding from premises or being based on evidence—and indeed in most cases actually preceding any evidence that would normally be thought to support them. But there is no point in adopting any of these ways of speaking. Whatever you call them, Hume's problem simply does not arise for guesses. Nor is there anywhere else that it can arise in the lifetime of a scientific hypothesis, since once guessed the hypothesis remains in the body of science until it is expelled; and the logic of expulsion, of falsification, and of supersession, is entirely deductive. Conjectures are not inferences, and refutations are not inductive. No inductive inference is needed to put a hypothesis into science; no inductive inference is needed to keep it there; and no inductive inference is needed to prise it out.

When Salmon said that "Popper's deductivism remains pure; science has no predictive import" (1978, 12), his premiss was true but his conclusion false. As Popper has repeatedly stressed (1974b, Section 15; 1972/1979, 363f), science does have predictive import (or predictive content): the conjectures of which science is composed are universal and are highly informative with regard to both past and future. But this does not imply that they are inferred inductively (or in any other way). The fact is that the rule of inference that Salmon used here, but dropped by his 1981, "if [science] has predictive import, it must incorporate some form of ampliative inference" (*op. cit.*, 13), is an ampliative one, a non sequitur, and so invalid. Of course, the predictions that science makes are not justified or warranted by the evidence; they are not reliable predictions. As Hume showed, this they cannot be. Why do they need to be? As I have emphasized in 1.1 and shall emphasize again in *g*. below, and in 3.4, our aim is to make true predictions, not reliable ones. I cannot understand what is supposed to be rational about demanding ampliative rules of inference that lead to reliable predictions when there is an uncontested (and, I guess, incontestable) argument that shows that no such rules can do anything of the kind.

c. The Empirical Basis

The third objection concerned the empirical basis. It is commonly supposed that, because test statements cannot be conclusively established on the basis of experience, some kind of inductive reasoning is needed to obtain them; if, after all, they are purely conjectural, then "it is hard to see why we should ever be satisfied even momentarily to rest science upon them" (O'Hear 1980, 75). All refutations, the critics feel, would become wholly arbitrary—which "paves the way for any arbitrary system to set itself up as an 'empirical science' ".

This was (as O'Hear recognises) exactly the objection that Popper levelled at Neurath's freewheeling doctrine of protocol sentences, according to which "one is allowed . . . simply to 'delete' a protocol sentence if it is inconvenient" (1934/1959a, 97). As an objection to Popper's own proposals, however, it is fundamentally mistaken, because it fails to take account of the role that accepted test statements (basic statements) are called on to play. According to falsificationism, test statements are used not for conclusive refutation, for the certification of hypotheses as false, but for the classification of hypotheses as false. In consequence, the important thing about the test statements we decide to accept—that is, classify as true—is that they be true; the only complaint that can properly be directed against a test statement is that it is false. Had Popper been less squeamish about mentioning truth he would surely have said this explicitly in his 1934/1959a. Now anyone who suspects that a test statement has been misclassified is welcome to test it in its turn, to try to show that it has false consequences. If he is unable to do this—not unable to test it, but unable to refute it—then clearly he has given no reason why the test statement should not be retained, why it should not still be classified as true. A refutation may be conjectural, but that does not make it any the less worth having, or make it incorrect.

Watkins too attempts to turn back on Popper his own criticism of Neurath's approach, that it scuppers empiricism. But in contrast to the other authors cited, Watkins does not conclude that falsificationism needs a rule of inductive inference, only that Popper has failed to show how it can avoid adopting one. In 7.2 of his 1984 Watkins imputes to *Logik der Forschung* the doctrine that the testing of a scientific theory amounts to no more than the decision to accept a test statement, with which the theory is then confronted; and this, he says, is quite arbitrary, and implies that no theory is ever put properly to the test. Test statements themselves, according to Popper, are tested only by confrontation with other accepted test statements; and according to Watkins, this means that there is nothing happening but "a lengthening chain of derivations; no *tests* are being made" (*op. cit.*, 253); experience disappears from sight. It seems to have been forgotten here that just as experience can only motivate the decision to accept a test statement, and cannot replace it (this would be psychologism), so equally the decision to accept a test statement can only be motivated by an experience, and cannot replace it (this would be conventionalism). When Popper says that "[e]very test of a theory . . . must stop at some basic statement or other which we *decide to accept*" (1934/1959a, 104), he hardly means that the making of the decision is all that there is to a test. This is clear from the next sentence: "If we do not come to any decision, and do not accept some basic statement or other, then the test will have led nowhere." Popper doesn't say there will have been no test; there will have been a test, but an inconclusive one (in the everyday sense of the word). An observation, or a measurement, has been made. To round it off we must decide on a test statement describing what has happened. Larry Briskman has brilliantly summed up Popper's theory of the empirical basis thus: *Look before you leap!*

There is nothing arbitrary in the demand that accepted test statements be true; objectively true, that is, not just consistent with other test statements, not just conventionally true. And in some areas of experimental physics we are lucky to have specific procedures, or so we conjecture, for classifying them accurately. (In biology, and especially in psychology, we are less lucky.) Those who object to any of these procedures, all of which culminate in the making of a decision, and to that extent—like all decision making—contain some conventional elements (*loc. cit.*, 108f), are invited to say what they think is wrong with them.

d. The Repeatability of Tests

It should be said at once that, on its own, the fact that a hypothesis has passed a test provides no reason for supposing that it will pass a repetition of that test. Nor does its failing a test provide any reason for supposing that it will fail a repetition of that test. But failure does imply that the hypothesis is false.

Hesse tells us in the passage quoted above that if we do not make the 'inductive assumption' that a falsified generalization 'All *A* are *C*' will be falsified in similar tests, that is, the inductive assumption 'All future *A* are not-*C*', then nothing obstructs us from relying on 'All future *A* are *C*'. There is some unclarity here, and the expression 'inductive assumption', though commonly employed, is unfortunate; for after all, induction is usually thought of by its aficionados as a process of inference, not as a strategy for making assumptions. But Hesse cannot mean that the restricted generalization 'All future *A* are not-*C*', which says that the falsified generalization "will be falsified in similar circumstances", must be the conclusion of an inference; the mere performance of a valid (deductive, or, if there is such a thing, inductive) inference cannot make something that was rational no longer rational; inferences are supposed to do no more than make explicit what is already implicit (deductively, or if there is such a mode, inductively) in the premises. So it seems that she can mean only that 'All future *A* are not-*C*' must be assumed. Falsificationists need not quarrel much with this, though it is not ideally expressed. Indeed, it was stated clearly in Section 22 of *Logik der Forschung* that a theory is falsified "only if we discover a *reproducible effect* which refutes [it]"; that is, only "if a low-level empirical hypothesis which describes such an effect is proposed and corroborated" (Popper 1934/1959a, 86). But the falsifying hypothesis may be no more than statistical; in any case, it is, like any other assumption, just a conjecture, and there is nothing inductive about a conjecture. I have gone through all this in *b.*, and I do not propose to repeat it.

What might be intended, and sometimes is intended, by the phrase 'inductive assumption' is a metaphysical principle of the uniformity of nature, what Schlesinger calls "the principle of induction" and describes (rightly) as "*entirely unprovable*" (*loc. cit.*). But whatever the merits of such an a priori principle may be, and we found few in our discussion of these things in *a.*, it is clear that nothing so vast is needed in order to obtain a falsification. Schlesinger may think otherwise, but it is unnecessary to verify 'All *A* are not-*C*' in order to falsify 'All *A* are *C*'.

In conclusion, attention must be drawn to the staggering inappropriateness of Ayer's often repeated question "What would be the point of testing a hypothesis except to confirm it?". The plain answer, after all, is that we test a hypothesis in order to refute it. The question that Ayer should have asked, the question that his own question naturally prompts, may be put inelegantly thus: 'What would be the point of testing a hypothesis in order to confirm it?'; or, more succinctly, 'What is the point of confirming a hypothesis?'. Interestingly enough, it was Ayer himself who raised this question in his most excellent criticism (1957), discussed in Chapter 7 below, of Carnap's requirement of total evidence (1950, 211–13). I have never seen a satisfactory explanation of the value of confirmation. As noted in 1.1 above, the fact that a hypothesis is confirmed (but not verified) has no bearing on whether it is true or false. Or, to put the matter in the terminology of belief so beloved of confirmation theorists: confirmation may make a hypothesis more worthy of belief than it was before, but it does not tell us whether we should believe it. What do we gain either by seeking to confirm it or by actually confirming it? I confess that I have no idea. The general question is further discussed in 3.4, and more briefly in Chapters 5 and 6.

e.   The Law of Diminishing Returns

The fifth objection concerned the severity of tests, and had two aspects: whether, in the absence of inductive activity, there exist any severe tests (I here reformulate as a substantive issue what Hesse presents as a conceptual one), and whether severity declines as tests are repeated. In response to the first point, it must be said firmly that—as the quotation from Popper 1963 in the previous section indicates—it is background knowledge, not past experience, to which assessments of the severity of tests are relativized. Background knowledge is by no means restricted to reports of previous tests; it contains generalizations of many kinds. There is nothing in the least inductive about this, as stressed in d. Indeed much of background knowledge is all too likely to be unexamined prejudice and presumption; and it is only through a severe test of a hypothesis that we sometimes discover how inadequate parts of our background knowledge are. For example it was presumably part of background knowledge at the time of the Michelson/Morley experiment that the length of a body in motion is independent of its velocity; this hypothesis, previously uncriticized, and surely never obtained by any sort of inductive inference, was eventually recognized to be false. Induction is in no sense a necessary condition for the existence of severe tests.

Severity of testing is relative not to some privileged sector of our knowledge—the already well-tested part, say—but to all our knowledge that is not directly in dispute. I can see that for an inductivist, who hopes to learn something from a severe test that a hypothesis fails to fail, who expects the virtues of the hypothesis to be objectively augmented by its success, this is unsatisfactory; for severity surely should not depend on the whims of some haphazard accumulation of background knowledge. But for a falsificationist there is nothing amiss. The requirement to impose severe tests is only a methodological requirement; and once a test has been carried out, it is not of the slightest interest whether or not it was severe. For if the test is passed, then nothing happens to the hypothesis; and if it is failed, then the hypothesis is rejected independently of whether the test was severe. As Popper rightly observes, the more we test a hypothesis, the more difficult it becomes to find new tests to which to expose it, and accordingly we are driven to tests of lesser severity; that "the empirical content of a theory grows stale after some time is an attractive asset to Popper's theory" (Boon 1979, 3).

Now the objection, raised by O'Hear and by Musgrave, that a falsificationist approach to science cannot explain 'the law of diminishing returns' is concerned not with the availability of new tests, but with repetitions of tests already administered (Boon, *op. cit.*, 2). At first sight the objection may appear to be a good deal more troublesome. Why should the fact that a test has been done several times before make repetitions of it uninteresting? My reply is that the point of repeating a test is no more than to check (but not, of course, to confirm) that we have discovered a decently reproducible effect. This is obvious in the case of tests that actually refute the hypothesis under investigation. All we need to know, once we have obtained an item of counterevidence, is that the effect recorded was not a freak; provided that it is not, we have a proper falsification, and nothing is to be gained by repeating the test—after all, we know already that the hypothesis is false. Exactly the same is true for a test that the hypothesis passes—we do not learn more from the repetition of the test than that it can be repeated with the same result.

Occasionally we may find that the test cannot be repeated with the same result—that the first effect discovered is indeed occult. Much more rarely we may even be led to conclude that there is no repeatable effect, even a statistical one, in the offing. That this rarely happens in the natural sciences is of course a significant factor in their successful evolution. That is to say, the truth of 'Most apparatus eventually behaves uniformly' (if it is true) is an important ingredient in the success of the natural sciences. It is even possible that this rather loose generalization could be smartened up sufficiently to become testable, and to qualify as a scientific hypothesis (Worrall 1989b, 269, describes the statement that "[no] accepted, well and independently checked, low level observational generalisation [turns] . . . out subsequently to be falsified" as "an overwhelmingly confirmed empirical generalisation"). But nothing justifies, or could justify, our assuming this statement to be true. Nor, as I have emphasized in a., is there any need to assume it.

Musgrave's own way of dealing with his objection is to say that a hypothesis becomes refuted (or corroborated) only when, as he puts it, "sufficiently many" repetitions of the test have been performed for "warranted acceptance of a falsifying hypothesis . . . or of a corroborating hypothesis" (1975, 251). But "in declaring the acceptance of some experimental generalization, as opposed to its negation, sometimes rational after 'sufficiently many trials', Musgrave is proposing a thesis inductivist in everything but name" (Howson 1977, 145). Musgrave's explanation has also been attacked by Grünbaum, on the rather unsporting grounds that it "leaves unsolved for Popper the problem of 'how often is sufficiently often'" (1976a, 237). But as I see the matter, the anticipation of a reproducible effect is normally a component of background knowledge before we begin a test; and we repeat a few times any test we have performed, not further to test the substantive hypothesis but simply to check that the effect we have recorded is indeed one that can be repeated. That these remarks are not bluff can perhaps be appreciated by considering a case in which background knowledge contains no suggestion of repeatability. Such cases are not easily come by. But suppose that we are testing the hypothesis that a set of galley proofs is entirely free of misprints. No one who knows anything about galley proofs will expect all the galleys to be alike, or even that there will be much of a statistical effect. In fact, until the first misprint is spotted every new galley will be effectively quite as severe a test of the hypothesis as is the first one; there will be no diminishing returns. (Kuipers has shown, however, that no appeal to induction is needed to obtain a law of diminishing returns in some cases of "tests for which we do not have good reasons to expect uniform results"; see his 1983, 210.) Neither here nor in the typical case does any question arise of the acceptance of a universal hypothesis in the wake of a number of repetitions of the same test. What might happen is that we have to reject the claim (if we ever made it) that the result of the first test can be repeated. But otherwise nothing happens, and since nothing happens, it cannot be too important when we stop testing. Anyone who is disturbed by this is perfectly free to carry on with the tests.

f. *Goodman's Paradox*

Goodman's example of the competing claims of the hypotheses 'All emeralds are green' and 'All emeralds are grue' (where 'grue' means 'green if and only if first observed before the year 2001') "is as clear a refutation as can be desired that a hypothesis can be confirmed, and that the confirmatory force lies in its positive instances" (Feyerabend 1968, 252). For the emeralds that confirm the first hypothesis confirm the second; and, indeed, confirm any other hypothesis that says that all observed emeralds are green and something different about unobserved emeralds. It is often asserted that the example (or some similar example) works as well against the view that generalizations are confirmed by instantial statements, such as 'At space-time region k there is a green emerald'. It has been claimed too that 'All emeralds are grue' will pass all tests conducted before the year 2001 that are passed by 'All emeralds are green' (this is controversial—but let it go); and so they will be equally corroborated by the evidence. How, then, it is asked, can we falsificationists have any empirical reason for preferring one of these hypotheses to the other? And if we have no reason, which prediction about the first emerald examined after December 31, 2000, shall we endorse: that it is green or that it is grue? Hardly both, it is suggested, since these predictions contradict each other (at least, they do so in the presence of the supposition that there will be emeralds newly observed in the next millennium).

My answer to this really quite unrealistic question is given by a simple refinement of Popper's criterion of demarcation. That told us that a hypotheses must not be admitted to empirical science unless there are tests that can eliminate it (and its claim to be true); for the same reason, conflicting hypotheses must not be admitted to empirical science unless there are crucial tests that can eliminate at least one of them (and so eliminate the problem posed by their conflicting claims to be true). If we are not prepared to delay until 2001 a decision between 'All emeralds are green' and 'All emeralds are grue', therefore, we must not admit them both to science; unless, of course, some other (not properly empirical) method of adjudication is made available. There is no other way out of it if the hypotheses are at present empirically indistinguishable (as the two under consideration assuredly are). This is barely different from the proviso registered near the end of 1.2 about isolated predictions. A prediction of the outcome of a single trial on a chance device may be a testable conjecture. That does not make it a genuine component of empirical science.

But which of the hypotheses shall we admit and which one shall we bar? The answer in this case is all too obvious (though in other cases it might not be). From an inductivist perspective, no doubt, in which only reasoned applications for admission to the body of science are considered, the lionization of 'All emeralds are green' instead of its rival must seem capricious; "the admission of one hypothesis and the refusal to admit another [are] . . . an arbitrary matter, the luck of the draw so to speak" (Lehrer 1980, 137). But falsificationism need not concede this. There is of course no justification for welcoming 'All emeralds are green' and (in consequence) banishing 'All emeralds are grue'. Still, no justification is needed. Lack of justification is not another name for irrationality. Moreover, the question of why 'All emeralds are green' is taken seriously, in a way that 'All emeralds are grue' is not, has an innocuous factual answer: greenness is pertinent to living in the world, and grueness is not. This answer is expounded by Quine, for example, in 'Natural Kinds' (1969); but by agreeing with Quine that many of the categories that we employ have an evolutionary ancestry, I do not intend to endorse much of the fashionable ideology of natural kinds, which is displeasingly unexplanatory and, as Hattiangadi (1987, 74) observes, pre-Darwinian. Nor am I capitulating to epistemological naturalism. Although our interest in greenness and our lack of interest in grueness, and more generally our background knowledge, must be recognized to be products of a somewhat haphazard evolutionary process of conjectures and refutations (many of them operating at a biological and preconscious level), the present situation is ill described as completely arbitrary, or as "an historical accident" (the phrase, but not the sentiment, is due to Worrall [989b, 275]). There is nothing arbitrary about refutation. In short, no appeal to induction is needed in order to explain why 'All emeralds are green' is in, and 'All emeralds are grue' is out. Nothing justifies their respective positions either. Indeed, we may be wrong. That goes without saying. Emeralds may be grue.

The reason Goodman's challenge seems so perplexing to justificationists is that they know perfectly well that the hypothesis 'All emeralds are grue' is silly, and does not merit serious consideration. But though they know it, they cannot justify it. Worrall, for instance (*loc.*

cit.): "there is, of course, no doubt that science . . . does, *as a matter of fact*, sanction what . . . are the intuitively right predictions. . . . but . . . the question all along was not *whether* science sanctions the 'intuitively right' actions as rational, but *why, on what grounds*, it does so." I am afraid that this was "the question all along" only for those who continue to hope against hope. The difficulty entirely disappears once we discard the mistaken idea that empirical support is significant, and that it is empirical support that makes a hypothesis eligible for admission into the body of science.

The above way of dealing with Goodman's paradox is not, I suppose, very different from that advanced by Bartley (1981), who points out that it is because it solves no problem that 'All emeralds are grue' has never been considered by gemologists. I would qualify this only to the extent of saying that it solves no problem that is not equally well solved by 'All emeralds are green'; thus one at least of the two hypotheses is redundant, and may be excluded from consideration. It has to be conceded that this answer has not found favour, even with the sympathetic Shearmur (1984); and some, such as Worrall (*op. cit.*, 266) have been pained by my dismissal of 'All emeralds are grue' as plain silly. But I described it so because I wanted to stress that, in science as in everyday life, it is whether our hypotheses are true or false that matters, not whether they are empirically supported; and we all know that 'All emeralds are grue' is false. We should desist from arid sceptical puzzles.

To this extent I agree with Maxwell 1993, but I have no sympathy with his disqualification of unattractively artificial hypotheses by means of the romantic metaphysical creed that "the world is non-aberrant" (69). Unlike Maxwell (whose position is at heart just another version of fallibilist justificationism) I fully accept the possibility of more realistic examples of testable theories, empirically indistinguishable here and now but diverging in remote regions of space and time, that have identical claims to scientific status. Maxwell thinks that scientists never countenance such possibilities. But—to take one instructive case—it has often been suggested (for example, by Wheeler 1977, 9–15, especially the penultimate paragraph of Thesis 1) that all laws of physics, and Einstein's gravitational theory in particular, collapse in gravitational collapse. Unless it is inconsistent, of course, Einstein's theory cannot imply its own failure; and hence we have two rival theories presented to us—universal Einstein and aberrant Einstein. The falsificationist response to this (or any other better) example would be just the same: we must guess which of the competing theories is true, and do our best (in the fullness of space-time) to falsify that guess.

g. The Pragmatic Problem of Induction

The question, often seen as the nub of Goodman's paradox, of how rationally to make predictions about the future, brings us to the matter of practical action, and the use of science in technology. Can decisions on how to act be made rationally without appeal to some dreaded principle of induction? I shall show that they can be. It is not such a difficult job. Yet I am aware of the almost universal opinion that falsificationism, whatever its merits as a methodology of theoretical science, runs aground here. Even supposed allies of falsificationism, outright opponents of inductivism, as I have noted, often lose their nerve when it comes to the rational making of decisions.

Some parts of the criticism are extremely easy to dispose of. In particular, there is an immediate answer to Worrall's question why the prediction that Galileo's law will not be falsified at its next outing does not involve recourse to a principle of induction: the prediction involves recourse only to Galileo's law. A non-inductivist can happily concur too with Worrall's earlier pronouncement (*op. cit.*, 257): "Accepted science not only does, but should, inform our technological practice. If someone wants to build a bridge that will stand up tomorrow or a plane that will fly tomorrow . . . [he] should assume in particular that currently accepted low-level generalisations will continue to hold tomorrow". All that need be said here is that if an agent accepts some low-level generalizations, he accepts implicitly that they will continue to hold tomorrow. No further assumption is needed. The only credible explanation of Worrall's insistence on the need for a "principle of induction" is that he, like Watkins, distinguishes the requirements of rationality in technology, and more generally in practice, from those in theoretical science; so that, in addition to the prediction that Galileo's law will not be falsified, we need also a substantive principle that we may rationally "rely on . . . the further predictions of already well tested *low level* empirical generalisations" (269). The same idea, that different standards apply in the realms of theory and of practice, is hinted at, though not fully endorsed, by Zahar too: "In technology a decision can be termed rational only if it relates to considerations of reliability" (1983, 168). As I have observed, Worrall himself thinks that this principle of the reliability (not only the truth) of well-corroborated empirical generalizations has to be accepted dogmatically (272)—by which, I suppose, he means no more than that it must be accepted without justification. I disagree. I do not think we need to accept it at all.

It is worth comparing these remarks with what Popper says (some passages from his 1972 have been quoted in Chapter 1): the rational agent should act as if the best tested of the surviving hypotheses is true, but not because there are good reasons for thinking that it is true, or that its predictions are true, or that "it will in practice be a successful choice" (1972/1979, 22); or, that is to say, that we can rely on the decision we take. Popper rightly offers no direct answer to the justificationist question of why we should act in this way. The most that can be said in response to it is that there are no reasons for supposing that an unrefuted theory is not true. The hypothesis that has most successfully survived the critical debate is, we conjecture, our best source of true information about the world, and if we wish to act appropriately, there is little sense in ignoring this information; or, worse still, acting as if it were not true.

I think that this answer of Popper's is approximately correct, but that it needlessly lays itself open to the kind of justificationist criticism that we find in the work of Lakatos (especially his 1968, Section 3.3 and 1974, section II[b]) and those whom he influenced. To start with, an important point must be made about how scientific theories are used in technology. Contrary to what is often said (by Popper himself in the passages quoted in Chapter 1; and, for example, in the quotations in section 1.g from Feigl [1974/1981] and Niiniluoto and Tuomela [1973], agents do not, in practical affairs, act on the basis of scientific theories; they do not use scientific theories as a guide to action, or as a source of positive advice. It is ceaselessly suggested that bridge builders build bridges on the basis of their scientific knowledge; para-digmatically, Newtonian theory. But the idea that there is any one scientific theory, or even any identifiable portfolio of theories, that tells us how a bridge is to be built across the English Channel (for instance) quite ignores the trite logical point that universal statements proscribe but do not prescribe: as stressed in *The Poverty of Historicism,* "it is one of the most characteristic tasks of any technology to *point out what cannot be achieved*" (Popper 1944/1957, 61). Medawar rightly observes that from the constructional point of view "a television set . . . is not within the effective comprehension of any one mind" (1985, 72f); a point to be set alongside Hayek's comment, approvingly cited on page 64 of Popper *op. cit.,* that the knowledge needed to tackle truly social problems cannot be centralized (1935b, 210). But the same thing is true even of a humble can of baked beans. There is simply no such thing as the scientific knowledge required for the manufacture of a television set or a can of beans; no amount of science will tell us how to tackle the job—there may be countless ways that will succeed. All that science can do is to tell us what lines of approach we do well to avoid. That is, we do not rely on, or even apply, scientific hypotheses when we act; we exploit them. I should like to dub this insufficiently appreciated truism *the slavery theory of applied science.*

Applied science is more like history than it is like theoretical science, in that when we are interested in applications we take the laws of physics for granted and search for appropriate initial conditions. How do we do this? How else than by conjecturing that some conditions will be appropriate (to produce, say, enough lift to get an aircraft aloft) and trying to show that our conjecture is mistaken? This elimination of tentatively conjectured initial conditions is characteristically done in one of two ways: theoretically—we use our scientific knowledge to rule out possibilities (that is, we exploit our knowledge)—and empirically—we try the conditions out, and see what happens. The latter method is of course decisive, but it is not 'using science'; moreover, as any test pilot will remind you, it leaves a lot to be desired. What pure science does not do is to suggest any assemblage of initial conditions that would be efficacious in achieving the kinds of practical goals that we desire. It is the business of engineers to discover sets of initial conditions that typically yield the right results. They are not just sitting there in the physics textbooks! Efficient sets of initial conditions are often incorporated later in instruments such as potentiometers, watches, and refrigerators, or in other artefacts such as bottles of aspirin or nail-varnish remover. The behaviour of each of these human products is described by a collection of well-corroborated empirical generalizations; and, except in degenerate cases (such as consulting a watch in order to find out what the time is), these generalizations themselves have to be exploited, rather than simply applied, when there is a practical problem to be solved with their assistance.

Some of this, I think, Worrall recognizes when he insists that we "rely on . . . the further predictions of already well tested *low level* empirical generalisations" (269). But the explanation for this is not, as both he and Lakatos (1969, 393) believe, that the humble empirical generalizations are more reliable than the grand scientific theories that soar above them. Reliability is not at issue; and it would be unfortunate if it were, given Hume's demonstration that all generalizations are utterly unreliable. If there is a difference it is that the aristocrats do not know how to work, and can be exploited only by experts. Instrumental generalizations are more approachable. Nonetheless, as Mill perceptively noted for the case of social engineering, they rarely suffice (1843/1961, Book VI, Chapter VI, Section 1, 572):

> It is not necessary even to the perfection of a science that the corresponding art should possess universal, or even general rules. The phenomena of society might not only be completely dependent on known causes, but the mode of action of all those causes might be reducible to laws of considerable simplicity, and yet no two cases might admit of being treated in precisely the same manner. So great might be the variety of circumstances on which the results in different cases depend, that the art might not have a single general precept to give, except that of watching the circumstances of the particular case, and adapting our measures to the effects which, according to the principles of the science, result from those circumstances. But although, in so complicated a class of subjects, it is impossible to lay down practical maxims of universal application, it does not follow that the phenomena do not conform to universal laws.

Once this point is properly appreciated, it becomes quite clear that a discussion of the way in which theoretical science informs practical action should focus not on the hypotheses that are consulted but on the proposals that we adopt and the decisions that we take. (See also Popper 1974b, Section 14, 1025f.) Suppose that we have some practical task ahead of us. Then there may be a number of proposals made concerning how we may best carry it out. I suggest that the rational thing to do is to follow that proposal that best survives the most searching criticism that can be directed against it. And I challenge anyone who disagrees with this suggestion to give a single example where he thinks that the advice given would be bad advice and would lead to failure. His argument, if he has one, will clearly constitute a serious criticism of the very proposal that is at issue, and so will lead to its not counting after all as the proposal that best survives criticism. The suggested rule of rational action here being advocated is, I concede, almost vacuous; like the rationality principle (Popper 1967/1983) in social and historical explanation, it is a kind of zero-method.

Doubtless things are often a good deal messier than this. In complex situations we shall rarely have a proposal before us that does not suffer from some disadvantage. But this means only that the making of a rational decision is a messier business, and not as obviously rational as it is in the simplest case. In some problem situations, in contrast, there will be several conflicting proposals that have stood up equally satisfactorily to criticism. But only if the criticism has been unforgivably perfunctory need this *embarras de choix* embarrass the critical rationalist. In many settings, such as that in which Buridan placed his ass, there are numerous equally effective ways of accomplishing the same task. All that rationality requires is that we avoid those proposals that do not survive critical scrutiny of their effectiveness. It is no criticism of a proposal's effectiveness to say that that effectiveness remains unauthenticated. That is a criticism of the proposal's credentials, of interest only to justificationists; but it is not a criticism of the proposal's effectiveness. Having no reason to think that a proposal will work is no reason to think that it will not work.

This answer involves no recourse to any principle of induction. There is no suggestion that the proposal that best survives criticism is reliable, or that we have any reason to expect it to be successful. We have no such reason; but as Briskman put it, mere "doubt is not criticism" (1977, 521). And the lack of any reason to believe in its success does not mean that the proposal will not be successful. Nor is it assumed at any point that the proposals that best survive criticism are in general those that turn out to be successful. Even if there were any need for this assumption, which there is not, we know far too much to be inveigled into believing it.

The answer coincides, in effect, with that given by Popper. For the proposal that best survives criticism will be one that is in accord with our best corroborated theories. To see this, note that the fewer lines of criticism we are prepared to tolerate, the more proposals will survive critical scrutiny, and in consequence the more will escape disqualification. If we are nervous about applying our theoretical knowledge critically, then many pretty dotty proposals will get through; if we are sufficiently finicky, and insist boringly on the possible falsehood of all our knowledge, then no proposal whatever will get rejected. In particular, those proposals that we all agree to be sensible will survive. But what is important is that these proposals will still survive when we wield our theoretical knowledge in a highly critical way. They might not survive, of course, if we were to use falsified or superseded hypotheses as ammunition; but this scarcely matters, since these hypotheses (unlike the unfalsified ones) cannot stand up to countercriticism; in any case we could scarcely regard the resulting decisions as scientifically informed. In other words, when all the critical activity is over, many proposals, including those that everyone agrees to be sensible, those that give the illusion of being based on the assumption that our best-corroborated hypotheses are true, will still be there unvanquished. Our decisions are not instructed by theoretical science, but selected.

Two points are crucial to this account of the rationality of making decisions that are scientifically informed. The first is that rational decision making (rational decision-making, that is, not rational-decision making) favours proposals that have survived criticism, not proposals that we have reason to expect to be successful (nor do we have reason to expect failure, for that would be a damaging criticism). The second crucial point is that the best we can do in the way of criticism is to deploy the most advanced theoretical knowledge that we have. That, remember, is what we conjecture to be true. Of course, we can be less zealous, and criticize more mildly. That will not disqualify the proposals that would survive harsher criticism, nor make them less appropriate. But it will inevitably compromise the rationality of the decision-making process.

In real life, I admit, there are numerous occasions when so little relevant scientific knowledge is available that proper criticism of competing proposals is impossible. Should I be inoculated against typhoid before my next trip to the tropics? Science, as opposed to uninhibited conjecture, is of little use here; what will matter is the cleanliness of the food that I eat and of the water that I drink, and there are all kinds of reasons why no scientific prediction of these factors can be made. (Even if there are statistical laws available, they will have little or no bearing on what is going to happen to me.) It is usually thought that in circumstances such as these (decision making under uncertainty) there is no option but to resort to judgments of subjective probability, or to considerations of utility, or (in Bayesian prescriptions) both. The unpleasantness of typhoid far outweighs the unpleasantness of its vaccine; and so if there is more than faint probability that I shall be exposed to poor drinking water I should be inoculated. This sounds like good common sense, but I nonetheless demur from the explanation. I am not averse to taking utilities into account, but, as already stated more than once above, I cannot see how recourse to probabilities can help. I want my action to be appropriate to what is going to happen (whether or not I know what that is), not to what is likely to happen. What I want to do is to maximize (or, more realistically, to maintain above a certain level) my actual utility, not my expected utility. It is for situations like this that insurance policies exist. But the real answer here is that in complex circumstances, and in all those where there is a limited quantity of scientific knowledge, the aim of the rational agent is not really to make the right decision (there may well be no such thing); it is to make his decision right. To put the matter less cryptically: making a decision is rarely the end of the affair; each decision has to be followed by innumerably many more, correcting and refining the initial one. If I decide not to be inoculated against typhoid, for example, I thereafter shun unwashed fruit. This piecemeal approach to decision making is hardly controversial, I hope, and is surely incontestable with regard to far-reaching decisions such as decisions to get married, beget children, seek induction into the priesthood, and so on. The rational agent is not the agent who gets things right, but the agent who is ready to correct his mistakes, as far as he is able, using, where possible, the scientific knowledge and expertise available to him. He calls on science not, I stress again, because he thinks that it is reliable (which it is not, whatever he may believe), but because he thinks that it is true.

How could the critics of falsificationism have got so wrong a point where, as one of them puts it, "there is really no room for any misapprehension"? Cohen (1978b) goes on to say of the view that no hypothesis has any degree of reliability that

> [some] of Popper's supporters, like David Miller, still seem to be content with this ascetic doctrine. Their advice to technologists is just to avoid proposals and theories which have not withstood the severest test available. That advice, so far as it goes, is scarcely objectionable. [Splendid.] But it remains wholly negative. Without deviating into inductivism, it cannot provide any rational encouragement to adopt those proposals and theories that *have* withstood appropriate tests, and it is this positive kind of encouragement that technology normally expects and derives from science.

A natural response to this expectation attributed by Cohen to technologists would be that (Cohen 1980b, 302)

> there are many things that a wise man might think it valuable to have (such as a method of squaring the circle) but nevertheless schools himself not to want since they are impossible for him to get.

But this would be to miss half the point. The fact is that this view, so uncritically championed by Cohen (and by almost all the other critics too), that it would at the very least be helpful to have "some way of appraising the evidential justification for relying on a given scientific hypothesis", is simply false. This point is developed, with acknowledgement to the *Meno,* in 3.4 below. I shall say no more here than that you cannot do better than act in the light of what is true. If you want to act appropriately to a situation, then true predictions of the effects of the competing proposals are all that you need. There is nothing to be gained from predictions that are reliable as well; after all, you will act just as you would have acted if guided by merely true predictions. (None of this applies without qualification to the Bayesian approach to decision making, criticized in Chapter 7 below, in whose deliberations truth plays no part.) As far as practical action is concerned, therefore, the long history of attempts to endow hypotheses and predictions with indices of reliability is a sleeveless errand of unbelievable sadness and futility. Granted, some measure of reliability might be of use if you wanted not merely to do what is right but to have some assurance that what you are doing is right. But then again, why should anyone think it necessary to have any such qualified assurance? I know of no answer that does not simply repeat at this point the very thing that is at issue.

It is not so much the view that we can do nothing without justification or reliability that has misled the critics, as the view that we can do anything with it.

h. Methodology and Verisimilitude

Very little need be said in response to Lakatos's objection that falsificationism needs an inductive principle in order to forge some connection between the methodological rules it proposes and truth (or verisimilitude). For there is a perfectly straightforward non-inductive connection between falsificationism and truth: namely that a hypothesis that contradicts a true test statement is false, whilst one that, so far as we know, does not, may be, for all we know, true. There are, I do not deny, difficulties in generalizing this elementary logical point from truth to verisimilitude. But the reason for this is plain enough—we still do not have any adequate theory of verisimilitude. (See Chapter 10 below.) What is clear is that just as experience enables us to dispose of the claims of hypotheses to be true, so too it should enable us to dispose of their claims to be truthlike or to exceed their rivals in truthlikeness. As long as k's claim to be closer to the truth than h is is open to observational or experimental refutation, but is not actually refuted, k may be preferred to h as a better approximation to the truth. This does not mean that we have any reason to suppose that k is closer to the truth than h is—only that we have no reason to suppose that it is not. Quite obviously, there is nothing resembling induction here. It is not suggested for a moment, even conjecturally, that if one follows "Popper's methodological appraisals" (or those of anyone else) "one has a better chance to get nearer to the Truth than otherwise". It is really very misleading to suggest that "Popper . . . once hoped to demonstrate that, by following the tenets of falsificationism, scientists would be led to produce theories of greater verisimilitude" (Caldwell 1991, 4; italics suppressed); and equally misleading to suggest that this hope had to be given up because of "technical problems in defining verisimilitude" (*loc. cit.*). The very idea that "there can exist in any science methodological rules the mere adoption of which can hasten its progress" was repudiated years ago by Klappholz and Agassi (1959, 74), a paper favourably cited by Caldwell. The few genuine methodological rules that exist remind us of things we should not do (for example, we should not resort to ad hoc adjustments in order to shield our theories from falsification), but give no indication of what we should do instead.

What may be suggested from time to time, perhaps only as an intuitive judgment, is that one scientific hypothesis is closer to the truth than is another. But this is something that, eventually, it is up to the facts to decide. For further discussion of the testability of statements of verisimilitude, and the unsettled prospects for a satisfactory theory, see Chapter 10 below.

i. The Miracle Argument

The final objection, for all its technical character, can also be quickly and painlessly disposed of. The plain fact is that excellent agreement with a host of precise predictions is by no means a monopoly of truthlike hypotheses; on the contrary, every hypothesis that is strong enough to make many predictions at all makes many very accurate predictions, however wild it may be intuitively. A beautiful example of how misleading intuitive judgements can be in this respect is given by Stewart (1989, 207f), under the title 'Two-edged Sword'. It follows from the work of Feigenbaum that all chaotic models of a very wide class yield exactly the same numerical predictions for two experimental quantities that describe the development of turbulence in dynamical systems. As Stewart remarks, empirical confirmation of the predictions of these values from a particular model could (in the absence of awareness of the universality property) naively be interpreted as powerful evidence in favour of the chosen model. That would be a bad mistake. In this case, admittedly, the evidence obtained is by no means improbable given other models, and the premises of the miracle argument presumably do not hold. But the point is not that the miracle argument is invalid as a piece of probabilistic reasoning (a version of it is spelt out in the technical note below), even though its conclusion may be much weaker than some scientific realists have hoped (see Musgrave 1988 and Worrall 1989 for further discussion of this point), but that its premises may hold very much less commonly than they are (on intuitive grounds) claimed to hold. In any event, as the argument of Chapter 11 below demonstrates, mere predictive accuracy simply cannot be taken as a mark of truthlikeness or nearness to the truth. This does not mean that we may not explain the good performance of a hypothesis in stringent tests by the conjecture that it is approximately true; and, indeed, the joint success of many incompatible alternatives is open to explanation in the same way. But why we should be so lucky in our search for truthlike hypotheses does not seem to me to be itself open to rational explanation. In a sense, it is almost unbelievable that we have done so well. That our theories are near the truth does not diminish the awesomeness of our success. Nor should it blind us to the fact that we still have much to learn.

Note however that even if the empirical success of a theory did raise the probability that that theory was near the truth, we would not have an inductive effect on our hands. It would be a probabilistic effect; but, as argued at length in Popper and Miller 1983 and 1987, there is really nothing inductive about probabilistic support. (A few details of this argument are included in the technical notes at the end of Chapter 3.) What would be properly inductive would be a rule of inference that let us infer approximate truth from empirical success. But a realist who is also a falsificationist will have no use for any such rule. I can be a realist without thinking that realism is justified or needs to be justified; without, in consequence, being affected by the argument of Laudan (1981/1984) against some of the more extravagant claims made for the miracle argument. Anti-realism is almost always a by-product of justificationism, and loses all its appeal when that blundering approach to knowledge is forsworn. Despite Popper's concession, that is, there seems to me to be no room for a whiff of inductivism here. That "reality, though unknown, is in some respects similar to what science tells us" (as Popper puts it) is, as we have seen in a. above, a consequence of what science tells us. It is not an additional obligation that science need at any stage incur.

2.3. Conclusion

The solution to the problem of induction is that induction, whatever it is, cannot be justified, but it can be dispensed with. There are non-inductive methods for exploring the world, equally resistant to justification, that can be articulated in advance of knowing what the world is like.

Falsificationism proposes a simple general method for undertaking the task of uncovering the universe's uniformities. But unlike all traditional and almost all modern forms of inductivism, it is not justificationist. It pursues not certain truth or justified truth, but plain unvarnished truth. Why do the criticisms listed above that falsificationism is besmirched with inductive elements recur, time and time again? Of course falsificationism is inductivist if all that means is that it proposes that we propose universal hypotheses for test. But what is intended by the critics, of course, is that falsificationism is unable to justify (in whole or in part) its role in the search for truth. I agree. But is it inadequate to its task? Those who still think so are challenged to say which of its recommendations or injunctions, its methodological rules, are unsuited to the business of sorting out what is true about the world from what is false. Is the use of observational and experimental reports not conducive to accurate classification? Is classical logic thought to lead us astray? Is the criterion of demarcation a bad way of avoiding irrevocable error? I do not think that any of these or other main ideas of falsificationism is thought by the critics actually to be objectionable; for all these ideas are components of any healthy methodology for discovering the truth. There may be better methods. Methods can improve just as instruments can improve. Towards a better methodology of science science itself will lead the way.

Only a justificationist could believe that in order to advance in science, or to act in a scientifically informed way, we need to invest ourselves with some general principle of induction or of the uniformity of nature, rather than content ourselves with the restricted (but testable) uniformities stated by scientific laws. Justificationists too usually realize that, even if some grander and more general maxim were available, it would not improve matters. There is no respect in which "the course of nature continues always uniformly the same" or "Actions that are ultimately guided by hypotheses that are *well corroborated* have the best chance of being successful" (Watkins 1984, 343) or "currently accepted low-level generalisations will continue to hold tomorrow" (Worrall 1989b, 257) are easier to justify than the corroborated generalizations that they are supposedly guaranteeing. The discussions of the pragmatic problem of induction by Watkins, Worrall, and many others reveal all too nakedly the instability and relentless futility of the justificationist addiction. Hooked on justification, the junkie finds himself impelled to demand more and more; his entreaties become progressively more and more outrageous and extravagant. Higher and higher level principles are injected into the system in the desolate hope that they can provide some justification for those lower down. But they do nothing of the kind. Far from satisfying the craving, they simply make it worse. It is high time that philosophers kicked the habit. Cold turkey is recommended.*

Technical Note

Let h be some hypothesis, say Einstein's theory of gravitation, and h^* be the hypothesis that h is near the truth (or nearer the truth than its predecessors). Let e be the evidence. We assume that $P(h^*) = \delta$ and that $P(e, \text{not-}h^*) = \epsilon$, both small numbers. Since e is meant to be the result of a test of h (and presumably also of h^*), we may assume also that $P(e, h^*)$ is near 1, say equal to $1 - \xi$. It is easily shown by Bayes's theorem that

$$P(h^*, e) = \frac{(1 - \xi)\delta}{(1 - \xi)\delta + \epsilon(1 - \delta)},$$

which will be quite high provided that $\epsilon(1 - \delta)/(1 - \xi)$ is much smaller than δ. This will hold, in particular, when ϵ is much smaller than δ.

It is far from clear that this condition will be met in practice. Given the enormous content of h, we should expect δ to be very small, indeed 0. The value of ϵ is, to say the least, enigmatic.

---

*Cold turkey: the plain unvarnished truth; sudden withdrawal of narcotics (*Chambers 20th Century Dictionary*, New Edition 1983, 246).

3

A Critique of
Good Reasons

in memory of Grover Maxwell, 1918–1981

Grover Maxwell saw with great clarity how little real justification there is in the world, and how paltry are our prospects for making any more. His efforts in confirmation theory (especially 1974, 1975) were, quite properly, devoted more to explaining why scientific hypotheses appear to be confirmable than to the vain task of showing that they are confirmable. I should like to think that the theses of this paper, that good reasons are unobtainable, that they are unusable, and that they are unnecessary, would therefore have incurred the warmth of his approval without forfeiting the heat of his disagreement. The issue I shall broach is a critical one for rationalism. I don't know how much better my treatment of it might have been had it had the benefit of Grover's inimitable brand of rational criticism.

3.1. Introduction: The Constitution of Rationalism

Despite everything that its title may suggest, this paper is intended as a defence of rationalism, not an attack on it. Not for the first time rationalism, like Kant in the face of Fichte, is in need of protection more from its supposed friends than from its acknowledged enemies. For many contemporary rationalists make such large claims on behalf of this fundamentally modest position as to lay it wide open to all the criticisms that irrationalists are eager to supply. My wish here is to issue a prospectus of the goods that rationalism is honestly and genuinely in a position to deliver. Everything on offer in the catalogue is currently in stock, and will remain in stock for as long as it is not found to be in any way defective.

In a sentence my thesis is this: rationality is concerned with reason, with argument, but not with reasons. Although there are such things as good arguments, and it is these that the rationalist strives to provide, there are no such things as good reasons; that is, sufficient or even partly sufficient favourable (or positive) reasons for accepting a hypothesis rather than rejecting it, or for rejecting it rather than accepting it, or for implementing a policy, or for not doing so. Indeed, one of the finest feats of the use of reason has been the unmasking of what pose as good reasons as pawns in a hollow and fraudulent imposture. Yet the illusion persists that the rational person is the person who can supply good reasons in favour of what he thinks; that the rational agent is the agent who can supply good reasons in favour of the decisions that he takes. All this is quite wrong. Reasons exist, no doubt, at least in the subjective sense, but not good ones; for there is nothing from which good reasons can be constituted, and from nothing they cannot be constituted. But this does not imply that rationality is impossible, either in intellectual affairs or in practice. Reason has a job to do in every sphere; reasons, poor things, have not.

It is generally recognized that in the wake of his onslaught on induction Hume lapsed into a form of irrationalism. His reaction is commonly judged to be irrationalistic because he contended that we cannot have any good reasons for any predictions that we may make; even such banal ones as predictions about tomorrow's sunrise. But it is not this step that is the disappointing one to the critical rationalist; what is so disheartening is Hume's further lurch into the disastrous doctrine that there is therefore no self-respecting role for reason itself. "Reason is, and ought only to be the slave of the passions", Hume concluded (1739, Book II, Part III, Section III, 415), a view that some philosophers still seem capable of finding comfort in. Hume's conclusion, that passions control reason, is as unfortunate, and as misguided, as his unstated premiss that reasons are the stuff of reason. This premiss is manifestly one that almost all rationalist philosophers still take for granted. They are wrong to do so.

One of the more tendentious philosophical books of the last decade or so, Stove's *Popper and After: Four Modern Irrationalists* (1982), though intellectually frivolous, brings these matters to a head. The book propounds the thesis that Karl Popper, who is from both an intellectual and a moral perspective the most distinguished modern champion of rationality, is covertly an irrationalist who dares not disclose the daftness of his own views and accordingly seeks to conceal their full force by linguistic legerdemain. Stove's book opens with this salvo (3):

Much more is known now than was known fifty years ago, and much more was known then than in 1580. So there has been a great accumulation or growth of knowledge in the last four hundred years.

This is an extremely well-known fact, which I will refer to as (A). A philosopher, in particular, who did not know it, would be uncommonly ignorant. So a writer whose position inclined him to deny (A), or even made him at all reluctant to admit it, would almost inevitably seem, to the philosophers who read him, to be maintaining something extremely implausible. Such a writer must make that impression, in fact, unless the way he writes effectively disguises the implausibility of his suggestion that (A) is false.

Popper, Kuhn, Lakatos, and Feyerabend are all writers whose position inclines them to deny (A), or at least makes them more or less reluctant to admit it.

Virtually the whole of Stove's case against Popper's rationalism (and the somewhat less transparent rationalism of the other accused) derives from the misconstruction enunciated in these first few lines. It is a misconstruction, and a bad one. For although it may be, as Stove suggests, implausible that (A) is false, (A) is false when it is understood in the sense in which Stove intends it; that is, as meaning that we have any more well-established knowledge than we had in 1580. Unquestionable it may be that we have more scientific knowledge now than we had four hundred years ago, but this scientific knowledge is not and never will be well established, or well founded, or well confirmed, or based on good reasons, or anything remotely like that. As Popper has been foremost in maintaining, what we call scientific knowledge simply is not knowledge in the philosophers' sense of being justified, or supported with good reasons. If it is knowledge at all, it is conjectural knowledge. Perhaps critical rationalists (as Popper and those who follow him describe themselves) are guilty of changing the meaning of the word 'knowledge' (Stove *op. cit.*, 14; Haack 1979, 311). Yet those who fancy that they have been loyal to the traditional sense of the word (tradition must not be presumed to stretch back to Descartes, it seems, let alone to Plato) have, in order to accommodate awkward discoveries, been forced to change its reference instead. This is incontestably the case with regard to the term 'scientific knowledge', whose extension once encompassed theories like Kepler's, Newton's, and Dalton's. Now, as linguistic purists are pleased to wield the term, it refers to nothing at all. Whatever else scientific knowledge is, it is not justified.

No one who understands critical rationalism could imagine that critical rationalists have been secretive about this; it is one of our principal contentions that human knowledge is always unjustified and unjustifiable, and one of our principal problems to explain how nonetheless it can develop and improve, and how scepticism can be kept at bay. See the presentation of the wonderful quotation from Xenophanes in Popper 1963, 25–27; Chapter 5 of Bartley 1962/1984; and Popper 1983, Part I, Chapter I, Section 2.1, 18–22. We recognize that scientific knowledge is most often not true either (though sometimes, we conjecture, it is), and great efforts have been made, in the name of verisimilitude, to explain how scientific progress is yet possible. No one acquainted with Popper 1963, Chapter 10; 1972/1979, Chapters 2 and 9, or 1966, Addendum 1 to Volume II, could question critical rationalists' devotion to truth and their refusal to blur the distinction between truth, which can be attained but is very rarely recognized, and justified truth (or certain truth or probable truth), which can be recognized but is very rarely attained. Yet the back cover of Stove *op. cit.* implicates Popper in "the prevailing view that scientific knowledge is never true (nor even probable), and never false (or even improbable)". And this judgement must have been the main philosophical stimulant of the absurd accusation of Theocharis and Psimopoulos 1987 that the impoverishment of British science in the 1980s was due in large part to the cheapening by philosophers of "the concepts of objectivity and truth". Finally, it is worth repeating the point made at the end of 1.2 above (and also with reference to Hayek under 2.2.g) that scientific knowledge belongs to no one, and exists in no one mind. In brief, scientific knowledge is usually unjustified untrue unbelief, all that the official view (as expounded, for instance, in Ayer 1956, but too frequently to need detailed citation) says that it ought not to be.

These issues are displeasingly verbal. I am prepared to concede, if I have to, that science is not knowledge; but I would then stress that science is interesting (or some of it is), and knowledge isn't. Maddening as their practice is, I am resigned to the usage of Stove and many others (for example, Salmon 1981; Newton-Smith 1981, Chapter 3; Lieberson 1982, 1983; Black 1983, 20; Zahar 1983, 168; Watkins 1984, 58f; and Musgrave 1991, 26f) that restricts the word 'rationalist' to those who think that good reasons of some kind or other need to be supplied if we are to have knowledge or science, or to make rational decisions. In that sense Popper is an irrationalist; so too are Bartley 1962, Briskman 1983, and others; and so am I, I am glad to say. In our view the renunciation of good reasons is an unavoidable consequence of the use of reason. We consequently renounce them.

3.2. Three Independent Theses

In this chapter I shall be putting forward three theses, and discussing them. It will be obvious why I do not say that I shall argue for these theses; for I am not going to argue for them, which would be an impossible undertaking. They are true nonetheless, I think.

My first thesis, a traditional and wholly admirable sceptical one, is that good reasons do not exist; it is impossible to furnish a good reason in favour of any thesis or action whatever.

My second thesis is that even if, contrary to the first thesis, good reasons were obtainable, they would serve no useful purpose; they would be quite redundant for anything sensible we wanted to do, intellectually or practically.

My third thesis, which is Popper's, is that good reasons are also unnecessary, in that we can think and act rationally entirely without them; rational arguments, on the other hand, are indispensable.

It is worth being clear from the start that these theses, especially the second and the third, are independent of each other. If you are on your own in Chicago, and need to be in Minneapolis within half an hour, then a sedan chair may or may not be obtainable; even if it is, it will be of no use; and there is no conveyance that will be of any more use. Something may be unavailable and unavailing, that is to say, but it does not follow that any adequate substitute can be provided. On the other hand, if you have half a day for your journey to Minneapolis, then a car will be of considerable use but is not crucial; a plane, an omnibus, or even a train, would get you there in time. Sedan chairs continue to be uncalled for, but now they are a good deal more futile than many other things. Good reasons are the sedan chairs of rational thought and action.

3.3. The Unobtainability of Good Reasons

The underlying argument of this section is a tediously familiar one, known under the title of the fallacy of begging the question or *petitio principii*. We start, for simplicity, with conclusive reasons or, as I shall call them in this chapter, *sufficient reasons*; this is in order to protect the discussion from any presumption that there is anything conclusive about the conclusions of arguments or inferences. Suppose that e is offered as a sufficient reason in favour of the proposition h. Then e will fail to establish or prove h unless e logically implies h. But if e logically implies h, h will not actually have been proved to be true, even though it may have been validly derived from e; for the derivation rests on an assumption, namely e, that itself asserts the truth of h (and perhaps more). As a proof of the truth of h the argument will be shamelessly circular. In an attempt to justify h the very question at issue will have been begged, and a rational person who recognizes the deductive connection between e and h and is not inclined to assent to h is not going to be inclined to assent to e either; or as Peirce and Ladd-Franklin (1901) put it:

> To accuse a man of begging the question is in reality a plea which virtually admits that his reasoning is good. Its only fault [the reasoning, not the plea] is that it assumes as a premiss what no intelligent man who doubted the conclusion could know to be true.

According to Hamblin 1970, 33, it was Peter of Spain, and more lately J.N. Keynes, who drew the distinction needed here between the fallaciousness of proofs and the fallaciousness of derivations. In the extreme case the reason e may be the same as the hypothesis h. That h implies h is uncontroversial enough; that h provides a reason in favour of h is nonsense. It is denied even by Stove (*op. cit.*, 68), who nonetheless maintains that sufficient reasons are sometimes to be had. In a technical note at the end of the chapter I show briefly why on Stove's own premisses (which I do not challenge) this is impossible.

　　To put the matter bluntly: just as the argument from 'It is necessary that if e then h' to 'e makes h necessary' (or to 'if e then h is necessary') is fallacious, and widely recognized to be so, so too is the argument from 'There is a sufficient reason to believe that if e then h' to 'e is a sufficient reason to believe h' (or to 'if e then there is a sufficient reason to believe h'). An argument may be fallacious, however, without its conclusion's being false. Yet if there is no sufficient reason to believe e, which may after all be known to be false, the conclusion of the above argument will be false. If h is in doubt in the absence of e, then there is no possibility for e to extinguish this doubt if it is itself in doubt. In this way every sufficient reason demands the earlier delivery of some other sufficient reason, in the presence of which it becomes otiose; in effect, it requires the earlier delivery of itself. There is no known way of stemming this kind of infinite regress. In lay terms it is a vicious circle, but circles are infinite regresses.

In the same way that the argument outlined above is fallacious, so is the argument from '*e* logically implies *h*' to '*e* is a sufficient reason to believe *h*'. This formulation succeeds, I hope, in making it plain that we can avoid any assumption that there are sufficient reasons in favour of the view that what we call the laws of deductive logic are valid, that there exists a justification of deduction. I insist that I make no such assumption, even though what we call the laws of deductive logic are indeed valid, as far as I am aware. I follow Bartley 1977, 469, in the opinion that the Tortoise's challenge to Achilles in Carroll 1895 is simply met. As long as a rule of inference is valid, the conclusion of an argument embodying it is true provided that the premisses are; there is no need for us to know, or to assume, that the argument is valid in order for it to be valid. Whether he appreciates what he is doing or not, the Tortoise commits himself to the truth of the conclusion of Achilles's inference once he commits himself to the truth of his premisses. Indeed, this is another way of stating the point about *petitio principii*: a valid inference, though able to reveal much, cannot justify anything that is already asserted by the premisses.

It is not often appreciated that there are two different arguments here, which happen to coalesce when what is under attack is proof; that is to say, sufficient favourable reasons. One is the standard sceptical argument against sufficiency or certainty: any inference reckoning to yield a sufficient reason for its conclusion needs premisses (as any inference does), and if the premisses are not proved, the conclusion is not proved. An infinite regress is immediately initiated. The second argument is invoked in the charge of *petitio principii*: the premisses of a deductive inference are not a reason of any kind for the conclusion—even an insufficient reason; but if there are reasons for the premisses, then these reasons are reasons for the conclusion too, and the need for the inference disappears. When the target is sufficient favourable reasons, or proofs, for which deductive inferences are (it is generally allowed) necessary, it is not easy to distinguish the effects of these two lines of argument. Both appear to say that the premisses of the inference cannot provide a sufficient reason in favour of the conclusion. I wish to dispel this illusion of identity.

First, a brief historical remark. The doctrine that every deductively valid argument is question begging as a proof or demonstration, even though there is no hint of fallacy when it is evaluated as a derivation or inference, is sometimes attributed (in Berkson 1990b, 98, for example) to Sextus Empiricus. But this attribution may not be quite safe; Sextus (who is not usually thought of as an original thinker) did discuss the kind of infinite regress argument inherited from the later Sceptics (*Outlines of Pyrrhonism,* Book I, Chapter XV), but this is by no means the same thing. I can find nothing in the report in J. Barnes 1990, Chapter 2, on Sextus's use of the mode of infinite regression, that impinges on the fallacy of *petitio principii.* At any event, it seems wiser to ascribe the doctrine that all valid arguments are question begging to Mill (1843, Book II, Chapters II–III), who discussed the issue in the special case of the syllogism (in particular the syllogism in Barbara).

The salient difference that I see is that the sceptical argument is directed at sufficient reasons, while the charge of *petitio principii* is directed at favourable reasons. If this is right, then we should expect the sceptical argument to be as effective against a sufficient negative reason as it is against a positive one. Indeed, it seems to be: there are no disproofs either. Equally we should expect the charge of *petitio principii* to be effective against a deductive inference that purports only to give a good (but insufficient) reason for its conclusion. This too seems right: as noted in the last paragraph but one, the premisses of a deductive inference are not a reason of any kind for the conclusion; *h-&-e* is no kind of reason for *h*. All valid arguments beg the question. I shall maintain in 3.5, however, that the charge of *petitio principii* is impotent against unfavourable or negative reasons—that is, against critical arguments. What force, we may ask, would the traditional sceptic have against insufficient reasons? This turns out to be a question that needs careful attention.

A switch of focus from sufficient to insufficient reasons is in any case overdue. For scarcely anyone these days thinks that any factual hypothesis can be established beyond doubt, or that sufficient reasons can be given for thinking any hypothesis to be true (and those who do think these things are mistaken). Currently popular opinion regards all hypotheses as potentially open to doubt, even those of evidence and testimony; perhaps even those of logic. Not all hypotheses are equally in doubt, however, and what differentiates them are the different insufficient reasons that can be rallied in their support. An assault on good reasons must take issue with good reasons in this sense. Although not everyone has an attitude to *petitio principii* that is as brisk and forthright as my own (compare, for example, the responses of Jackson [1984], Mackenzie [1985], Walton [1991]), there are few, I think, who are still inclined to take seriously the possibility of sufficient reasons.

It is not necessary to defend the view that the great majority of rationalists, though eschewing sufficient reasons, do take good reasons seriously, and indeed seem incapable of discussing rationality in any other terms. But the dire situation of sufficient reasons must prompt a question about the possibility of merely good reasons. Do they exist?

Since sufficient reasons are assembled from sufficient premises and valid inferences, there appear to be two places at which insufficiency could enter: in the premises, which would be probable, or plausible, or something along those lines, but not certain; or in the inference, which would be plausible or probable, or something along those lines, but not valid. Of course, both sources of insufficiency may be present. Indeed, the second is impossible on its own if traditional scepticism is right; whatever the inference is like, its premises are not certain. What about a deductive inference with uncertain premises? I have claimed above that such an inference continues to beg the question, and to provide no reason whatever for its conclusion. It is wrong to suppose, as does Berkson, that "the vulnerability to the infinite regress argument comes from the demand for conclusiveness [sufficiency], and not from any particular logical structure of argument" (1990a, 86; see also his 1979, 299). Whether Berkson is right about traditional scepticism is, as mentioned four paragraphs ago, a more subtle question. If only deductive inferences are admitted it is hard to banish the thought that the premises cannot be less open to question than the conclusion is, so that the sceptic's argument remains successful. But this, I think, is to take covert advantage of the fact that deductively valid inferences are also question-begging, and to miss what is distinctive about the sceptical challenge. Let us therefore restrict our attention to inferences that do not presume to be deductively valid.

A limpid analogy here is between complete certainty and immaculate cleanliness. Without cleaning agents that are perfectly clean we cannot expect to get anything else perfectly clean. (For simplicity I assume a continuity theory of matter. Remember that this is an illustrative analogy only.) Still, very high levels of cleanliness are possible, and clean rooms in physics laboratories achieve them. Might it not be possible, despite Lewis's much-quoted remark that "[if] anything is to be probable then something must be certain" (1946, 186), to dispel uncertainty

very successfully, but not completely, using premisses that are themselves uncertain? The general opinion, amongst Bayesians and others
who incline to see everything in probabilistic terms, and amongst others
too, is that this is extremely possible. (The original Bayesian attempt to
show how uncertain evidence is processed is Jeffrey 1965/1983,
Chapter 11, but there are others. One aspect of Jeffrey's theory is taken
up for discussion in Chapter 7 below.) For my part, I do not intend to
go into this question, and am ready to concede an affirmative answer.
But I want to maintain that low levels of uncertainty, however attained,
are unable to underpin good reasons. The objection is the same as
before—that in each case the question is being begged.

It is perhaps as well to make plain at this point that I do not
undertake to refute every form of the doctrine that good reasons can
exist. After all, they can always be defined into existence (what I take to
be an instance of such stipulative rationalism is discussed in 6.3 below).
It would help if I had a better idea of what good reasons are supposed to
be good for. But this has never been revealed to me. I think, on the
contrary, that they are useless, a point elaborated on in the next section.
In the absence of any keen intuition about the intended role of these
good reasons, I can only follow the lead of those who most enthusiasti-
cally advocate their acquisition.

Just as sufficient reasons, were there any, would be sustained by the
relation of logical implication or logical consequence, so good but
insufficient reasons presumably need a relation of partial implication or
partial consequence. There is no genuine agreement concerning what
properties such a relation possesses (Salmon 1969), but it will suffice to
think of it as being spelt out in probabilistic terms. Let us suppose that
we are granted a measure P taking values between 0 and 1 inclusive such
that if e implies h then $P(h|e) = 1$; and if e is consistent and implies not-h
then $P(h|e) = 0$. (If e is inconsistent it implies both h and not-h.) One
obvious measure of the degree of partial implication from e to h is $P(h|e)$
itself; it may be taken as measuring the proportion of states of affairs
where e holds that are also states of affairs where h holds. We may say that
e partially implies h to the degree $P(h|e)$, so that P takes the maximum
value 1 if e does imply h and the minimum value 0 if e implies not-h.
There are largely familiar considerations that suggest that this relation is
not necessarily the best one, especially if it is to be used as a basis for
judgements of what are and what are not good reasons. (An agent who
knows scarcely anything about me may assign a very high probability,
perhaps around 365/366, to the proposition that my birthday is not

August 19th. But it would be odd to say that he had any good reason for this judgement.) What is at first sight a less perverse measure of the degree of partial implication from e to h is what is known as the measure of support or relevance

$$s(h, e) = P(h|e) - P(h),$$

where $P(h) = P(h|taut)$ for any tautology *taut*. It is clear that for any fixed h the term $s(h, e)$ increases along with $P(h|e)$. It may be negative too, if $P(h|e)$ is less than $P(h)$, and zero if e is probabilistically independent of h. For any fixed h, the function s takes the maximum and minimum values under the same conditions as P. There are therefore at least two possible ways of construing the claim that e partially (but not fully) implies h: first, that

$$0 < P(h|e) < 1,$$

and secondly, with greater refinement, that

$$0 < P(h) < P(h|e) < 1.$$

Note that on the assumption that $0 < P(e)$ it follows from the inequality $P(h) < P(h|e)$ that $0 < P(h)$. Salmon has discussed these two relations of partial implication at some length. Oddly enough, he concludes that they are "radically incongruous with one another" (1969, 70), even though one of them is simply a refinement of the other.

What is often seen as the main advantage of the measure s is this: except in cases where some initial probabilities take the extreme values 0 or 1, a hypothesis h is always supported by its logical consequences; that is, if h logically implies e, then $s(h, e)$ is positive. This is not easily interpreted as meaning that e is a good reason for h (how good depending on how large s is), since e would then be a positive reason, perhaps not a strong one, but some kind of reason, for every hypothesis that implied it. Can we say that e provides a good reason for h if $s(h, e)$, the support that h receives from e, is high? It seems to me that the argument from 'e strongly supports h' to 'e is a good reason for h', or even to 'There is a good reason for h' or 'If e there is a good reason for h', is just as starkly fallacious as the argument from 'e logically implies h' to 'e is a sufficient reason for h'. And when h and e are the same, or e implies h, the conclusion is as false as it was a few paragraphs ago: e is not a good reason for itself or for any of its logical consequences. In this special case at least it looks like the business of begging the question all over again.

Now this objection is so obvious that it may well seem that I must have missed the point. The obvious objection that will be raised to my objection is that although the argument from '*e* supports *h*' to '*e* is a good reason for *h*' is evidently invalid if the premiss includes the case in which *e* implies *h*, it may not be so if this extreme case is excluded. Whenever a reason is sought for a hypothesis *h*, it may be allowed, some question has to be begged, some premiss has to be granted; but this involves no fallacy unless *h* is, or is part of, the premiss. The fallacy is one of begging the question, not begging a question. If *e* is brought forward as a good but insufficient reason for *h*, then it is the question of *e* that must be dealt with before the question of *h*, not the question of *h* itself. Though there is admittedly still a potential regress to worry about, no real circle is involved here.

My retort to this retort, with which I have a measure of sympathy, since it does recognize some of the force of the fallacy of *petitio principii*, is to demonstrate that it is not correct. The point can be made in the simple case in which *h* implies *e*, when, as we have seen, $s(h, e)$ is positive. For this positive support is entirely to be explained by the fact that *e* is part of the content of *h*. As already emphasized, *e* cannot provide any sort of reason for itself; citing *e* as a reason for *h* therefore begs part of the question at issue, though not all of it unless *e* and *h* are identical. What is left? That is to say, what part of the content of *h* goes wholly beyond *e*, what part of the question raised by *h* remains wholly unbegged by *e*? If we wish to encapsulate this excess content in a single proposition, then the only promising candidate is the material conditional *h*-if-*e* (or if-*e*-then-*h*). This conditional has no factual content in common with *e*, and together with *e* it is exactly equivalent to *h*. Now it turns out quite generally that a proposition *e* provides positive support for a proposition *k* only when *e* and *k* overlap in content. Except in trivial cases *e* actually countersupports *k* (that is, $s[k, e]$ is negative) if *e* and *k* have no content in common. It follows that *e* countersupports any part of *h* that goes wholly beyond *e*, in particular *h*-if-*e*. (Some details of all these matters, which are by no means entirely uncontroversial, are gathered together in technical note *b*. at the end of this chapter.)

In the light of these results the doctrine that *e* can provide a good reason in favour of *h* when it offers it positive probabilistic support seems indefensible. The evidence *e* provides no reason at all for part of *h*; and the residue of *h* is countersupported. Hence what is promoted as a good reason for *h* is at the same time a good reason against some of the consequences of *h*. Where this leaves the connection between rationality and truth defeats me. (For some further defective properties of the measure *s*, see Carnap 1950, 366–397, especially 397; or the excellent brief report in Section 3 of Salmon 1975.) There are measures of partial implication that avoid this fault, for example the measure *q* of deductive dependence of Miller and Popper 1986, under which *e* is neither favourable nor unfavourable to *h-if-e*, but neutral. (See Technical Note *b*. below.) But there is no variety of good reason that I know of that permits *e* to speak in favour of a hypothesis with which it has no common consequences; and none at all that can allow it to speak in favour of its own consequences.

Black, in a brief discussion of the metaphor of support (1983, 37–40), notes that *e* can support *h* (in the sense used above) even if *e* itself is quite unsupported; and from this appears to conclude that *e* can provide a good reason in favour of *h* as long as it is true. But if this were really so, there would be a good reason in favour of almost any consistent hypothesis, since every hypothesis has many true consequences. Black gives examples also of propositions that offer each other mutual support. (Witnesses in court often support each others' stories.) I freely admit, of course, that support, and even good reasons, can exist in such senses. What I deny is that they have any relevance to the theory of rationality.

This concludes the discussion of my first thesis; good reasons do not exist, and indeed cannot do so. This applies just as much to reasons that do not aim to be sufficient as it does to those that do. As I have stressed, I cannot prove this thesis. But it is surely up to those who think that good reasons exist to explain what truth emerges from allowing any proposition to act as one of the judges in its own cause.

3.4. The Unusability of Good Reasons

My second thesis is that even if good reasons could be obtained, they would not serve any purpose. This is perhaps not quite correct with regard to sufficient reasons. For if indeed we could establish the truth of a proposition *h* beyond any possibility of doubt, if we could render its truth absolutely certain (by whatever means), then we could classify it as true. As I have explained at length in 1.1, the classification of truths and falsehoods is the primary business of intellectual investigation.

As far as I know, nothing is gained by also certifying hypotheses as true. Still less is there any point in providing them with showerproof, rather than waterproof, guarantees. Insufficient reasons alone have no power to authorize the classification of any hypothesis among the truths. This is, after all, not to say much more than that an insufficient reason for h is not a sufficient reason for h. The natural question to ask me at this point is why I ask so much of insufficient reasons. Is it not enough to ask of an insufficient reason for h that it provide a partial authorization for the classification of h as true? The answer is no. For a partial authorization again fails to issue an answer to the question of whether we should classify h as true—a question that, as far as I can see, we shall eventually have to answer without calling on any authorization or reason. For no reason at all is needed for classifying a hypothesis as true; and an insufficient reason doesn't do anything that no reason at all doesn't do quite as well. If we wrongly classify h as true, our error will not be mitigated one bit by our having had a good but insufficient reason for the classification. And if we rightly classify h as true, our being in possession of a good reason will not embellish our achievement in any way—not, that is, unless we desire more than the classification of statements into the true and the false. (If we are interested in justice, we shall not require that justice be seen to be done, provided that it is done. Of course, matters would be somewhat different if we were interested also in seeing justice to be done. The most that we normally manage to achieve, it should be noted, is seeing that the procedures of justice are carried out correctly.)

Exactly the same considerations apply in the sphere of practice. Here, despite appearances, all that we actually need is a true prediction concerning how matters will turn out; there is not only no need for, but also no use for, a good reason to think that the prediction is true. A beautifully modulated, yet widely disregarded, passage in Plato's *Meno* makes just this point at 97A–D.

> *Socrates*   Let me explain. If someone knows the way to Larissa, or anywhere else you like, then when he goes there and takes others with him he will be a good and capable guide, you would agree?
>
> *Meno*   Of course.
>
> *Socrates*   But if a man judges correctly which is the road, though he has never been there and doesn't know it, will he not also guide others aright?

Meno Yes, he will.

Socrates And as long as he has a correct opinion on the points about which the other has knowledge, he will be just as good a guide, believing the truth but not knowing it.

Meno Just as good.

Socrates Therefore true opinion is as good a guide as knowledge for the purpose of acting rightly. . . .

Meno It seems so.

Socrates So right opinion is something no less useful than knowledge.

Meno Except that the man with knowledge will always be successful, and the man with right opinion only sometimes.

Socrates What? Will he not always be successful so long as he has the right opinion?

Meno That must be so, I suppose. In that case, I wonder why knowledge should be so much more prized than right opinion . . .

Meno's wonder is, in my judgement, well placed. But regrettably Socrates, who up to this point has argued unerringly (even to the extent of sorting out the muddle, implicit in Meno's penultimate speech, between hypotheses and policies that are guaranteed or reliable in the sense of working every time and those that are so in the sense of being accredited; that is, of being known to work every time), now slips into an apologetic defence of knowledge (rather than true belief), on the grounds that knowledge is stable, or tethered, whilst true opinion is evanescent and fleeting. "That is why knowledge is something more valuable than right opinion", he goes on (98A); but he does not hesitate to state again that "true opinion when it governs any course of action produces as good a result as knowledge" and that "for practical purposes right opinion is no less useful than knowledge" (98A, C). In the jargon of this paper, Socrates sees some purpose for good reasons—to transform truth into indelible truth, truth away from which you cannot be persuaded (it is clear that only sufficient or absolutely conclusive reasons would be equal to this task); but he sees no practical role for them. As far as practice is concerned Socrates is absolutely right. Knowledge does nothing.

I often ask those who continue to find this Socratic insight unacceptable the following question. *Which ticket would you prefer to draw in a sweepstake: the one bearing the favourite's name, or the one bearing the winner's?* The answer I receive is almost always that the winning ticket is, of course, the best one to draw; and an agent's preference in the abstract would be for this ticket. But I am forcefully reminded after a semicolon's pause that, until the race is over, no one can know which ticket is the winning ticket, and as a matter of tactics the rational agent therefore prefers the ticket most likely to win (or the ticket for which there is the best reason to expect victory). It is the 'therefore' here that takes my breath away. If the tactical preference for the most favoured ticket is not to be simply an underhand repudiation of the abstract preference for the winning ticket, then the agent must have conjectured that the ticket most likely to win actually will win; and he prefers the favourite not because it is the favourite but because he conjectures that it will be the winner, and evaluates the truth of that conjecture in the light of the evidence. Those who enthuse over probabilistic methods, or over the rational significance of good reasons, will claim, I don't doubt, that the cerebrations and investigations relevant to the conjecture that a hypothesis is true will also induce us to assess it as probable or reasonable. That may be so, but what matters is the conjectural classification of the hypothesis as true (which is independent of whether it is also assessed as probable or reasonable), not its assessment as probable or reasonable (which may well depend on its previous classification as true). The subjective feeling of being in possession of good reasons may exist. But as far as rational thought is concerned, evaluation in terms of good reasons is a pure epiphenomenon.

3.5. The Unnecessariness of Good Reasons

My third thesis is in most respects the most fundamental and challenging of the three. It is simply that rationality can operate perfectly well without recourse to good reasons; reason nowhere depends on reasons.

The kind of rationality I have in mind, admittedly a far cry from the aspirations of authoritarian rationalism, is usually called critical rationality or critical rationalism, to emphasize the critical role that criticism plays within it. Since justifications, partial or complete, or good reasons, sufficient or insufficient, are unattainable, it should not be demanded that we provide arguments, or reasons, or evidence, or grounds in favour of a hypothesis before we entertain it or classify it as true (or if such things are demanded, it should not exercise us if they are not forthcoming). If rationality is to be given anything to work on, we must be allowed to propose any hypothesis we like as a candidate for the truth. The only restriction should be not whether the hypothesis actually is true, but whether, if false, it can be overthrown and rejected. So the role that logic, or reason, plays in critical rationalism is a critical role. Logic, as Popper has said (1963, 64), is the organon of criticism, not of proof. The way science proceeds is by the proposal of conjectures, hypotheses; by the derivation from these conjectures of further propositions that can be tested by experience, or by other means (that doesn't really matter here); and, should the consequences derived turn out to be false, by the rejection of the original conjectures. Valid derivations transmit truth, as we all know. But they also retransmit falsity from the conclusion to at least one of the premises. It is in this way that criticism is the lifeblood of reason.

Let me make this point, elementary as it is, absolutely plain. Critical rationalism insists that we should unfailingly resort to logic as a way of criticizing the hypotheses we have before us. That is, critical rationalism insists that we pay attention to reason. No one who reads this paragraph with any kind of understanding can fancy, as do some of the authors previously mentioned, that critical rationalists, by abandoning the search for justification or reasons, have thereby abandoned also the rigours of rationality.

Now we come to the tricky bit. Isn't criticism just proof, or the provision of good reasons, turned on its head? Suppose that I conjecture *h* and derive from it *e*, which turns out to be false. It looks at first sight as though critical rationalism may have to say that not-*e* is a reason to reject *h*; or, putting the matter logically rather than methodologically, a reason against *h*. This seems hardly different from saying that not-*e* is a reason for not-*h*. Doesn't critical rationalism, that is to say, rely just as much on good reasons as does traditional rationalism? To revive the objection laboured a couple of sections ago, doesn't not-*e* beg the question of not-*h*, simply because it logically implies it? To be sure, there can be no good reason for not-*e* that is not at the same time a good reason for not-*h*. But unless we have such a good reason for not-*e*, or something suspiciously like it, it seems as though we shall not have one for not-*h* either. So the objection goes, we have no reason for not-h, no reason to reject h. Thus all rejection, falsification, and criticism, unless calling at some point on good reasons (negative or critical reasons, perhaps, but still reasons), are ultimately arbitrary.

There are two responses open to the critical rationalist here, one specious and superficial (though not wrong), the other simple and satisfying. The first response would be to observe that the question-beggary of critical reasons is by no means as scandalous as it is in the case of supporting or justifying reasons. Although not-e may beg the question of not-h, it does not beg the question at issue, namely h. What is so debilitating about the fallacy of begging the question, it may well be maintained, is not that to provide a justification of h we have to assume something already to be justified; but that we have to assume h itself already to be justified. No circularity is noticeable in the case of critical arguments (unlike that of purported partial justification, whose concealed circularity was uncovered in Section 3). Nor is the difference between justifying arguments and critical arguments simply "a trick of the propositional calculus" (Harré 1986, 46), but the most fundamental feature of the whole of logic and rational thought. (For further discussion see my 1994.) In critical arguments there admittedly is a regress (or, as it will transpire to be, a progress), which is doubtless deeply disturbing to those who crave for justification, but there is no circle.

This can be seen more readily if we consider a number of steps in the chain. If you wish to justify or prove h, you may cite h_1 as a reason for it. Called on to justify h_1, you subpoena h_2; then h_3; and so on. If each of these reasons does what it was summoned to do, then h, h_1, h_2, h_3, . . . will stand in the relation: h_3 implies h_2, which implies h_1, which implies h. It is obvious that h_3 begs the question of h quite as much as h_1 does, and for all the good it does it might just as well have been cited at the outset (though someone genuinely concerned about justification would more wisely have stuck at h). However far back we go along a chain of proofs, h will always be at issue. In contrast, once we advance a few steps along a critical chain, we shall find ourselves dealing with quite different questions. I propose h as a hypothesis. You criticize it by deriving consequence e, and proposing that not-e is true. I counterattack by criticizing your claim of not-e; I derive from it k, and claim that not-k is true. Now k could of course be not-h, in which case we shall have landed ourselves in a circle from which there may be no escape (if we are perverse enough). But k need not be not-h; it may be some other proposition, one that is quite independent of *h*. My assertion that not-*k* is true may therefore be a quite separate issue from the one we started from, whether *h* was true. Thus the dialectic of criticism, unlike the monologue of justification, is not committed to the endless iteration of the same point.

An unnaturally simple and schematic example may serve to make this evident. I propose *P*-&-*Q*. You derive *P* from it, and assert not-*P* (you may have carried out an empirical test). In my attempt to thwart you I derive *R*-if-*P* from this conjecture of yours, and assert not-(*R*-if-*P*), that is, *P*-&-not-*R*, to be true. To criticize this, you derive from it not-*R*, and claim that *R* is true. Here we have reached an issue quite different from the original one, though doubtless overlapping it in content (but this does not matter, since neither *P*-&-*Q* nor *R* is being cited as a reason for the other). We are no longer fighting over the same question (though that need not mean that the original issue of *P*-&-*Q* has been entirely forgotten). There are many examples of disputes in the empirical sciences that begin at a lofty level and fizzle out in altogether different and humbler territory—for example, in a test of whether passing tramcars cause significant vibrations in nearby laboratories. Moreover, as should be clear, we are not launched on to an infinite regress so much as on to an infinite progress. New problems are being posed, and old ones conjecturally solved. That the new problems typically form what Lakatos (1970/1978, Section 2[c]) pejoratively calls a degenerating problem shift is unimportant. Critical paths may be retraced. In the case of justification the infinite regress or circle is so damaging because without the first prop (which actually does not exist) all the others are worthless (a point apparently disputed by Black *loc. cit.*). But an infinite progress without a final goal is nothing like as serious. Of course, it has to be brought to a stop somewhere, and normally we shall have no real difficulty in finding somewhere to stop. But we do not need to have any reason for stopping where we do stop.

There is therefore a bit more to the famous "*asymmetry* between verifiability and falsifiability" than what "results from the logical form of universal statements" (Popper 1934/1959a, 41). All that Popper says there (and in 1983, Part I, Section 22) is correct, and anticipates without difficulty the obvious formal objection gleefully raised by such critics as Binns (1978, 33) and Schlesinger (1988, 171f), that whenever a hypothesis is falsified its negation is verified. But the truth is that falsifiability remains the central consideration even when unrestricted existential statements are in question. 'There exist neutrinos', for example, becomes discussable only when it is rendered into a falsifiable form—as it is by the specification of a [universal] recipe for producing and trapping neutrinos (Popper 1974b, 1038). Reports of tests have to be accepted, of course, but nothing is justified thereby. The question always remains open whether the test reports are correct.

To summarize this first response: critical arguments may beg questions, but they do not beg the same question at every step. There is inevitably a circle of justification. There is no critics' circle.

Although there is much that is right in this response, it smacks of casuistry. The trouble is that it takes as a problem, rather than as an irrelevance, the justificationist's contention that negative or critical reasons are possible only if we also have positive ones. This is an irrelevance to critical rationalism, since critical rationalism need have no more truck with negative reasons than it does with positive reasons. There are no negative reasons either; nor do we need them for rational thought and action. We do not need reasons against a hypothesis in order to classify it as false (nor against a course of conduct in order to classify it as foolish). All we need is a false consequence of it; not, I would stress, a reason to suppose that we have identified a false consequence; just a false consequence. It follows, I need hardly say, that we are in permanent peril of classifying statements incorrectly and of doing things wrongly. This, I am afraid, often happens; though even if it did not, we would still be in peril. It does not follow that we do not often get things right.

Many will fear that this abnegation of negative reasons as well will void science of its last vestiges of rationality and leave it at the whim of entirely arbitrary decision. What, it will be demanded, is to stop us from classifying as false, or true, any statement we like? If we need no reason for classifying *h* as false then it does begin to look as though all constraints (apart perhaps from logical consistency) are unbuttoned. This would doubtless be the case if we embraced also a coherence or relativistic theory of truth. (A drift to relativism is the most common outcome of the clash between the confusion of truth with justified truth and the realization that there can be no justified truth.) But no one who does science does embrace relativism (though some have been told that they do, and believe it). The aim of science is indeed truth, but this is something like correspondence to reality. And it is because we are aiming at truth that our classifications are not arbitrary, even though they may well be wrong. There is, to be sure, no reason why 'All swans are green' should not be classified as true. Still, this classification will be incorrect, and if we are as eager to sort out truth from falsity as we claim to be, we shall test the hypothesis, perhaps showing that it has consequences that contradict other statements that are classified as true. These other statements will not supply us with a reason for rejecting 'All swans are green'. But if we are concerned about truth we shall not persevere with them all (coherence is necessary for truth, but quite insufficient); we shall have to decide which of them to reject. We may do this wrongly.

It is obviously essential for all this that we should be prepared to classify statements as true or false without having a reason for so classifying them. It is not essential that we should always perform this classification correctly.

I wonder if there can exist a single rationalist or realist who thinks that the method of proposing criticizable theories and subjecting them to unstinting criticism is not, as it happens, an excellent (though admittedly always fallible and sometimes inefficient) method for sorting out truth from falsehood. The method, everyone agrees, works well on the whole, even if there is never any reason to believe that it might not be malfunctioning and leading the investigation in the wrong direction. Critics of critical rationalism should at least try to state clearly what it is that they are attacking. Is what is under fire the method that we propose (a method that, in some disguise or another, the critics usually advocate too), or is it our frank refusal to garnish our product with spurious and deceitful claims of reliability? The two are not the same, and to dress up a criticism of the second kind as one of the first kind is simply mischievous. In some critical comments there is perhaps an unspoken accusation of freeloading; critical rationalists coolly assert that the method of conjectures and refutations is a good one, but they are covertly taking advantage of all the hard work put in by epistemologists and confirmation theorists in pursuit of a justification of their method. I deny this completely. Not only is there almost nothing to show for all this hard work beyond the rediscovery of the fact, known at the outset, that justification (whether sufficient or insufficient) is impossible. Even if critical rationalism were to be justified, this justification would be of no interest to critical rationalism, which manages well without it. Truth and falsehood do get separated.

In sum, the critical rationalist's answer to the question 'Why do you think that *h* is true?' or 'Why do you think that action *a* should be performed?' will be 'Why not?'. But this is an invitation to cite some disadvantage of *h* or *a*, not to marshal reasons against it or in favour of some alternative. Sometimes, as in arithmetic or in the law, where some principles are agreed by all parties to a dispute to be inviolate, a demand for a good reason, or for a proof, may simply be a disguised request for a derivation. That must be allowed (and we must not be misled by the terminology). But in all cases where the premisses are deemed not to need support (whether, as in the law, because they are given, or because, as some have maintained of statements of perceptual experience, they are given), their consequences do not need support either, so there is no sense in which the derivation does more than elicit the consequences. Usually, however, a challenge to provide good reasons has to be met firmly with an appeal to the challenger to instance a drawback in what is being proposed. Only those whose interest in truth is quite frivolous will want to take advantage of not being asked for a good reason to back up their objection. Truth is unquestionably above and beyond human responsibility. We must be thankful that in addition it seems fully capable of withstanding the indignities imposed by the excesses of human irresponsibility.

Technical Notes

a.   *Stove on Conclusive Reasons*

Stove writes that "*h* cannot be part of a reason to believe *h*" (1983, 68), and more generally "[that] the premisses of an argument entail the conclusion is *not* enough to make them a reason to believe it" (67). Moreover "any stronger [proposition] . . . than *e* will be logically equivalent to *e*-&-*f* for some *f*, and if *e* cannot be [a reason or] even part of a reason to believe *h*, then evidently no proposition logically equivalent to *e*-&-*f* can be so either" (72). Write *h* in place of *e* here, and use the principle stated at the beginning of the paragraph. We may conclude that no proposition equivalent to *h*-&-*f* can be a reason, or part of a reason, to believe *h*; which means that no proposition logically stronger than *h* provides a reason in favour of *h*.

This I can very easily assent to, for it looks like a straightforward recognition of the fallacy of *petitio principii*. Since *e* could hardly be a conclusive reason for *h* if it did not imply *h*, we can conclude further that there can never exist conclusive (sufficient) reasons. Yet Stove thinks that conclusive reasons are sometimes attainable. He says, for example, that " 'All men are mortal and Socrates is a man' is an absolutely conclusive reason to believe 'Socrates is mortal' " (77). I do not understand how anyone can be expected to believe this.

b. Excess Content Is Countersupported

All propositions have among their consequences all logical truths, but it is not difficult to show that in classical logic the two propositions h-if-e and h-or-e share no other logical consequences. Moreover, h-if-e and h-or-e are together just sufficient to imply h. (The assumption in 3.3 that h implies e is here relaxed.) It is therefore eminently reasonable to identify the excess content that h has over e with the conditional h-if-e (Hudson 1974). Every consequence of h that is not a consequence of h-if-e shares content with e; the only part of the content of h where e does not beg the question is h-if-e.

That the evidence e will normally countersupport the conditional h-if-e may without great hardship be demonstrated ab initio from the axioms of the calculus of probability. But a much simpler proof can be given from the following two propositions (one of which was adverted to earlier): first, that $s(h, e)$ is non-negative if h implies e,

$$\text{if } h \text{ implies } e, \text{ then } P(h|taut) \leq P(h|e);$$

second, a weak form of the law of complementation

$$\text{if } P(x|z) < P(y|w), \text{ then } P(\text{not-}y|w) < P(\text{not-}x|z).$$

For by writing e-&-not-h (which is equivalent to the negation of h-if-e) for h in the first of these, we obtain

$$P(e\text{-\&-not-}h|taut) \leq P(e\text{-\&-not-}h|e),$$

and hence by the second,

$$P(h\text{-if-}e|e) \leq P(h\text{-if-}e).$$

Identity will hold only if either $P(e) = 1$ or $P(h|e) = 1$, neither of them cases of interest if h is a hypothesis and e some evidence called to support it. We may conclude that in all interesting cases the excess content h-if-e is undermined or countersupported by the evidence e.

To my mind the function s not only does not measure inductive support, it does not measure anything. What it may have been confused with at an intuitive level is the function $q(h|e)$ of deductive dependence of Miller and Popper 1986 (see note c. in 10.4 below), which measures (crudely) the part of the content of h that is covered by e. It may be shown that whenever h and k have no content in common (beyond logical truths) then $q(h|k) = 0$. There is no support, but there is no countersupport either. One of the nice properties of q is that it increases monotonically with the second argument; it cannot go down as evidence piles up. This has the consequence that $q(h|e)$ can be positive when h is contradicted by e, so it certainly does not measure what has been called degree of rational belief. The similarity between the measure q and what Keynes (1921, Chapter VI) called "the weight of arguments" has been remarked on by Hilpinen (1970, especially 106, 111).

Further details of the proof that excess content is countersupported, and the controversy that envelops it, may be found in Popper and Miller 1983 and 1987, and also in our 1984 reply to the comments of Levi 1984, Jeffrey 1984, and Good 1984. Recent discussions include Zwirn and Zwirn 1989 (see also my rejoinder, 1990b); Mura 1990 and 1992, 71–87; Dubucs 1993, 89–91; Zwirn and Zwirn 1993.

4
Comprehensively Critical Rationalism: An Assessment

in memory of Bill Bartley, 1934–1990

In this chapter I should like to pay tribute to Bill Bartley by saying something about his most enduring philosophical work, The Retreat to Commitment, *and in particular about the doctrine of* comprehensively critical rationalism *(CCR) expounded there. Although I was shocked and saddened by Bill's untimely death, I shall not allow this to deter me from forthrightly criticizing his views. This is the least I can do to honour his memory. Bartley himself pulled few punches in intellectual matters, and he did not expect restraint from his opponents.*

But I shall defend him too. Indeed, my thesis is that the main criticisms that have been levelled at CCR are wrong, and that the position can be sustained in something very much like the form in which it was originally put forward. What I shall not feel obliged to defend are Bartley's own responses to these criticisms, which seem to me often to be inadequate.

4.1. Bartley

Bartley—in the formal mode of speech, Professor William Warren Bartley, III—was born in Pittsburgh, and educated at Harvard and at the London School of Economics. In 1962 he submitted a Ph.D. thesis, *Limits of Rationality: A Critical Study of Some Logical Problems of Contemporary Pragmatism and Related Movements,* written under the supervision of Karl Popper, and published *The Retreat to Commitment* (1962). For a while Bartley was Lecturer in Logic in London, and he then took up an appointment at the Warburg Institute. He later held positions at the University of California at San Diego, the University of Pittsburgh, Gonville and Caius College Cambridge, and California State

University, Hayward. He was, I am told, an unusually able lecturer—I never heard him lecture—and his dedication was recognized in 1979 when he was named Outstanding Professor in the 19-campus California State University system.

Although *The Retreat to Commitment* is undoubtedly the book by which Bartley's purely philosophical achievement must be judged, he published much else, including many papers and reviews. Worth noting here are a short monograph *Morality and Religion* (1971); the biography *Werner Erhard: the Transformation of a Man* (1978); his edition of Lewis Carroll's *Symbolic Logic* (1977/1986), which made public for the first time the lost Part II of that work, a document that Bartley himself unearthed; and the extraordinary *Unfathomed Knowledge, Unmeasured Wealth* (1990), published shortly after he died. There is also, most importantly, his edition of the three volumes of Popper's long-delayed *Postscript* (Popper 1982a, 1982b, 1983). Since Bartley's criticism of early sections of this work seems to have been one of the factors that led to its non-appearance in the 1950s (when it was written) it is appropriate that it was he who undertook the labour of seeing it through the press, eventually in a greatly expanded form.

In 1973 Bartley published a fascinating and highly controversial biography *Wittgenstein* that led, he maintained, to attempts on the part of Wittgenstein's friends and followers to ostracize him from professional philosophical existence, at least in the United Kingdom. The trouble was not so much philosophical as Bartley's claims about Wittgenstein's private (or, it suggested, not-so-private) life; in brief his allegations of Wittgenstein's homosexuality, and of his penchant for visiting the Prater Park in Vienna in search of gratification. The book was described as "foul" by a close friend of Wittgenstein's and as "filth" by one of his relatives, yet the reviewer in *Christian Century* called it "chaste". The revised edition, published by Open Court in 1985, contains a spirited reply by the author to his critics. An evaluation of Bartley's accusations is to be found in the Appendix to Monk 1990.

In the last few years of his life Bartley took on three huge tasks: the editorship of *The Collected Works of F.A. Hayek*, and quasi-official biographies of both Hayek and Popper. In the course of this work he was appointed Senior Research Fellow at the Hoover Institution on War, Revolution, and Peace at Stanford, where the archives of both Hayek and Popper have been deposited. All three projects are very much unfinished; so unfinished that it is not, I think, possible to say how unfinished they are. Of the 22 projected volumes of the Hayek *Collected Works* only one had been published before Bartley died, and even now only a handful have appeared. The project is continuing. Volume XI, which will contain the edited correspondence between Hayek and Popper from 1937 to the present, is expected to be of particular interest to those concerned with liberal theory and with methodological problems in the social sciences.

Bartley had considerable personal charm. He could also be extremely difficult. He was not afraid of controversy; and indeed, did not seem to want to keep away from it. He was a man of courage and integrity. I remember being filled with admiration at the manner in which he spoke from the floor in response to Grünbaum's carping address on Popper and psychoanalysis at the LSE International Symposium in 1980. Much as I disagreed with Bill on many things, I felt always that he was a philosopher whom it was not wise to ignore.

4.2. Comprehensively Critical Rationalism

I now turn to CCR, which is the constructive kernel of Bartley's most original work, *The Retreat to Commitment*, published first in 1962, and in a revised and expanded edition in 1984. The book is meant as a contribution to "the theory of the open mind", as he calls it at the start of his preface; that is, to a study of the strengths and weaknesses of rationalism when opposed to theories of human thought that endorse dogmatism, dependence on faith (fideism), obeisance to authority, and sublime indifference (relativism, anything-goes) as ways of settling intellectual disputes. The problem posed by Bartley springs from an unresolved crisis of identity or integrity in contemporary rationalism that is highlighted when it is put alongside contemporary Protestant thought (the thought, in particular, of the theologians Barth, Niebuhr, and Tillich). It needs to be said that Bartley writes most fascinatingly about religion, and somehow succeeds in making the subject infinitely more interesting than it really is. According to Bartley, Protestantism is able to defend more or less with integrity its "retreat to commitment", its unargued faith, only because rationalism too, even at its best, has been ultimately forced to rest itself on faith—on faith in reason. Such an irrational (or, anyway, non-rational) dependence on reason is resorted to in Chapter 24 of Popper's *The Open Society and Its Enemies* (1945/ 1962), for example. "Why should we be ashamed of making an irrational commitment", Bartley saw Protestants as asking, "when rationalists do just the same?" In response, Bartley set out to provide a new theory of rationality—originally called comprehensively critical rationalism, later renamed pan-critical rationalism—that can resist this *tu quoque,* taking nothing on faith; it was to be a theory of rationality unashamedly able to live up to its own standards. We have all heard the story of the zealous but humble clergyman who ended an impassioned sermon with the words, *Don't do as I do, do as I say!* Bartley's problem was to develop a theory of rationality with no need for such humility.

Following Popper 1945, 229f, Bartley distinguishes two strains of rationalism in our culture, *comprehensive rationalism* and *critical rationalism,* the latter a mutant of the former (or perhaps developed by a process of hybridization). Comprehensive rationalism is the view, traditional in Western philosophy, that we should accept, or adopt, or believe in propositions only if they can be defended or supported or justified or proved by means of argument or experience. In some form or another it is to be found explicitly in Descartes, in Berkeley, in Hume, and elsewhere; implicitly almost everywhere. It is an easy prey to sceptical arguments, to the ever-looming infinite regress of justification or support, and to the charge of *petitio principii* treated at length in 3.3 above. Popper dismisses comprehensive rationalism as inconsistent; for, he claims, it cannot itself be supported by argument or experience, so by its own lights it should not be adopted. That is not formal inconsistency, more like what Austin (1962, 14) called infelicity; in any event, it is surely gravely damaging. Popper's own response is critical rationalism: a doctrine that may be summed up in the attitude: "I may be wrong and you may be right, and by an effort we may get nearer to the truth" (238). Critical rationalism, as he puts it, "recognizes certain limitations" (229), acknowledges that "no rational argument will have a rational effect on a man who does not want to adopt a rational attitude" (231), and is reluctantly prepared to put up with—but not disguise—the fact that "the fundamental rationalist attitude results from an (at least tentative) act of faith—from faith in reason" (*loc. cit.*). The text makes it plain that the "critical" in "critical rationalism" here means *self-critical,* modest, aware of its own shortcomings, unpretentious, and so forth.

A doctrine that is similar to this in partly excusing rationalism from its own requirements, but from a moral point of view starkly different, is also placed by Bartley under the banner of critical rationalism: this is the slippery, evasive, self-centred, complacent rationalism advocated by Ayer in his discussion (in his 1956) of the problem of induction. Ayer suggests (75) that the question of the rational status of induction cannot even be asked, since induction is constitutive of rationality. To wonder whether induction (or rationalism) is rational is as senseless, Ayer says, as to wonder whether the legal code is legal. The details, and the detailed criticism, of this position are of no real importance here. (See Section 2 of Salmon 1968.) All that we need record is that Ayer, like Popper, treats rationality as a domain with singularities; locations, that is, where normal standards do not fully apply. For a valuable further comment on this point see Watkins 1987, 270.

Bartley set out to rescue rationalism from this impasse; to find a formulation of rationalism that is free of singularities—so, like comprehensive rationalism, comprehensive—yet self-critical and rationally sustainable.

His solution is *comprehensively critical rationalism* (1962, 147f; 1984, 119): "A position may be held rationally without needing justification at all—*provided that it can be and is held open to criticism and survives severe examination.*"

This solution is underpinned by two crucial adjustments to the force of the word 'critical' in 'critical rationalism'. In Bartley's hands 'critical' came to mean not merely 'sensitive to criticism', but something like 'constituted by criticism'; in addition he forced us to see that the appropriate standards of criticism are not those that appeal to justification, partial or total, but those that appeal only to truth. In the tradition of (fallibilistic) eliminative induction, say, rational (justified) belief is the goal, and the critical method no more than a good way of reaching that goal. For Bartley, in contrast, it is the method of investigation itself, not its outcome, that is rational (or not rational); it is truth that is the goal. There can be little doubt that both the fundamental ideas here—*i.* the shift from epistemology to methodology, and *ii.* the clear separation of (positive) justification from (negative) criticism—have a common origin in Popper's solution of the problem of induction. But Bartley deserves much credit for having appreciated the decisive character of these oppositions, especially *ii.*, for having sharpened them, and for having exploited them in the solution of the problem of rationality.

Let me say a few words about each. *i.* That Popper's solution to the problem of induction is a methodological one is plain. Throughout his work, beginning with 1934/1959a, Section 5, Popper characterizes science by its method; a method of bold conjectures subjected to rigorous and severe testing. The method is dominated by the rule that we should not protect our hypotheses from falsification. This means far more than that we should be ready to take notice of falsification, to discard falsified hypotheses; it means that we must actively seek falsifications out. Except in the final chapter, 'Corroboration', Popper's emphasis in 1934/1959a is primarily on how we investigate hypotheses in our possession, and little on how their status is affected by the investigation. My own view is that, for as long as a hypothesis escapes refutation, its epistemological status is entirely unaffected. It is as good (or as bad) as the day that it was born. (See 1.2 above, and the concluding remarks of 6.3 below.) This opinion might seem to conflict with Bartley's remark quoted above; for it seems to suggest that a hypothesis may be held rationally only if it has survived severe criticism, whereas I say that it may be held rationally even if it has just this moment been conjectured; or, perhaps, like so many commonsense ideas, if it has been around for ages and nobody has noticed it. But a later passage indicates that Bartley thinks this too: "a comprehensively critical rationalist, like other men, holds countless unexamined presuppositions and assumptions, many of which may be false. His rationality consists in his willingness to submit these to critical consideration when he discovers them or when they are pointed out to him" (1962, 151; 1984, 121). The reason that, usually, it is rationally indefensible to adopt or hold a criticizable position that has not survived severe criticism should be obvious: either it has wilted under criticism, and so deserves to be eliminated, or it has not been put to the test. The latter is not the fault of the hypothesis, but a fault in the way it has been processed; the irrationality lies only in our laziness, in our failure to be sufficiently critical. (It is worth noting that the only idea from traditional epistemology that is required here is the idea of truth.)

As for *ii.*, Bartley did well to emphasize that the traditional way to criticize a hypothesis in philosophy—and still the way most favoured in many quarters—is to show that it is unfounded or unproved, or to show that a supposed argument for it is invalid or unsound. Criticism, that is to say, has traditionally meant criticism for failing to meet the appropriate epistemological standard of justification. In science things proceed rather differently (though there are occasional lapses). In science a hypothesis is criticized by showing that it fails to meet the standard of truth; in short, that it is false. It is this sort of criticism, criticism of content, stressed also by Popper (1960/1963, 22), that is central to CCR.

In preparation for a consideration of the criticisms that have been launched at it, we should now state CCR more carefully. Like critical rationalism, like comprehensive rationalism too, CCR is a system of methodological rules rather than a body of doctrine, as the following summary of the "new conception of rationalist identity" indicates (Bartley 1962, 146; 1984, 118):

> The new framework permits a rationalist to be characterized as one who is willing to entertain any position and holds *all* his positions, including his most fundamental standards, goals, and decisions, and his basic philosophical position itself, open to criticism; one who protects nothing from criticism by justifying it irrationally; one who never cuts off an argument by resorting to faith or irrational commitment to justify some belief that has been under severe critical fire; one who is committed, attached, addicted, to no position.

This is clearly not a mere redefinition of the word 'rationalist' but a set of directives, rules, advice, what have you, to the aspiring rationalist. To follow the rules it is, of course, not enough merely to announce a preparedness to take on would-be critics. Every position that a comprehensively critical rationalist holds must be so formulated that genuine criticism is a live option. That is, criticism must be possible, even if—because the position is in fact correct—no successful criticism exists.

There therefore seems to be built into CCR an assumption to the effect that *all positions are in principle potentially criticizable*. Bartley explains: "when I declare that all statements are criticizable, I mean that it is not necessary, in criticism, in order to avoid infinite regress, to declare a dogma that cannot be criticized . . . ; I mean that it is not necessary to mark off a special class of statements . . . [that] are not open to criticism" (1984, 222f). Let me call this the CCR-generalization. It is important to realize that this doctrine is not intended as a doctrine of pure logic, but one of methodology. (That does not ensure that it is not logically false! For criticism to this effect, see *d.* below.) Whatever the true extent of the CCR-generalization is, one consequence of it that CCR must uphold is that CCR itself is potentially criticizable.

4.3. Criticisms of CCR

There are still, I am fully aware, several obscurities in the above statement. But I hope that they may be resolved through a consideration of some of the objections that have been made to CCR so formulated. In this section I shall look at six types of objection: *a.* logical truths and other trivialities are uncriticizable, but almost everyone accepts them; *b.*

some doctrines (reinforced dogmatisms) are deliberately phrased so as to be uncriticizable; *c.* CCR is vacuously correct; *d.* CCR is paradoxical; *e.* CCR is committed to deductive logic, which is not criticizable; *f.* CCR does not tell us how to react to or take account of successful criticism.

a. Trivialities

It might be objected that logical truths, tautologies, and arithmetical identities, are not even potentially criticizable, since they are necessary truths. We know in advance, especially with the really elementary ones (such as 2 + 2 = 4) that no criticism can possibly defeat them. The same goes for the more extensive class of analytic truths, such as 'All bachelors are unmarried', and for a number of statements that are synthetic all right, but trivially true; for example, Watkins's "There exists at least one sentence written in English prior to the year two thousand that consists of precisely twenty-two words" (*op. cit.,* 274). Hence the CCR-generalization is incorrect.

There are a number of ways of answering this objection as it stands, though it will return later in e. in a more implacable form. Perhaps the first thing to note is that when we talk of the criticizability of a proposition, we must have in mind some particular formulation or class of closely related formulations. There is one sense, no doubt, in which necessary truths are not open to criticism, but it is not that that is intended here. Given this, the right response to the objection, I think, is to note that we have developed methods for checking the correctness of addition sums and logical derivations—even very trivial ones—and these we sometimes need; for example, if we are very young or very intoxicated. There is thus a perfectly clear sense in which I know how to criticize the statement that 2 + 2 = 4, even if, thanks to my age and abstemiousness, I rarely need to perform the check. To be sure, no criticism of 2 + 2 = 4 will be successful. But the CCR-generalization no more requires that all positions be successfully criticizable than Popper's demarcation criterion requires that all scientific hypotheses be successfully falsifiable. This response gives a hint, perhaps, of one of the advantages of systematization: we understand what a criticism of 2 + 2 = 4 or of a simple truth table would be like because we have a method that applies to all sums, or all truth tables. What is at issue here is not whether a particular sum is a consequence of Peano's axioms, or whether a particular statement is tautological, but whether it is true. The truth table method (together, in general, with contingent information) permits us to check whether any truthfunctional compound is true; not just tautologies and inconsistencies, but contingent ones as well.

As for trivial synthetic truths, their criticizability in principle is rather obvious. It may normally be insincere for me to talk about holding the statement 'I am more than three years old' open to criticism. But I can easily specify ways in which it could be criticized. (Think of the implications of the twin paradox in special relativity.) That is enough. The matter is even more straightforward for incontestably decidable synthetic statements such as the example offered by Watkins (which, unintentionally, I imagine, contains only 21 words); one way to criticize it would be to query whether it is correct English to write 'twenty-two' rather than 'twenty two' or to write 'the year two thousand' rather than 'the year 2000'. (It is not important whether these criticisms would be successful.)

Bartley claims also (*op. cit.*, 244) that examples of this kind bear no real relevance to the problem situation of CCR. There is no situation, he asserts (more or less), in which a would-be rationalist would feel obliged to compromise his rationalism and terminate an argument by refusing to consider criticism of 'I am more than three years old'. I am not so sure. It is all too easy to be exasperated by someone who denies the obvious. What is important is that the obvious may always be opened to investigation, and its very obviousness makes investigation easier.

b. Reinforced Dogmatisms

It might be objected that there exists another important class of statements that cannot seriously and sincerely be held open to criticism, what Popper (1963, 327) called reinforced dogmatisms. A reinforced dogmatism is a doctrine into whose statement is built a strategy for diffusing and defusing all kinds of criticism. If criticism is restricted to empirical criticism, then several well-known views (for example, some Freudian doctrines) have been accused of incorporating such strategies. To avoid needless controversy, let me suggest as a trivial example the hypothesis that physical objects go out of existence when no observer (person, animal, or instrument) is observing them. Skilfully formulated, this hypothesis simply cannot be tested (which is not to say that it cannot conflict with other—empirical—hypotheses). So it may not, Popper says, be introduced into empirical science.

It is much harder to give an example of a reinforced dogmatism so impregnable that it can deal automatically with all possible criticism.

The clearest example is perhaps a crude form of irrationalism: 'All argument is impotent and illusory'. Bartley, following Popper, takes this to be a consequence of the doctrine of metaphysical determinism (1984, 259), and discusses it briefly. He acknowledges that, as a rationalist, he might by rational argument be led to such a position. Is it a position that a comprehensively critical rationalist could adopt? In one respect, it is manifestly not. A rationalist is obliged to take argument seriously. On the other hand, there is no reason at all why a rationalist should not seriously discuss this doctrine, and argue about it. Such discussions go on all the time: witness Popper's extended discussion (in 1982a, Section 24, and also in Popper and Eccles 1977, 75–81) of the possible irrationalistic consequences of metaphysical determinism; and the endeavour known as the strong programme in the sociology of knowledge, which asserts—with not much plausibility, it must be said—that all intellectual activity, including itself, is socially determined.

There seems nothing at all wrong with this, provided that the rationalist allows, as Bartley certainly does (*loc. cit.*), that he might eventually have to admit that it is he who is in the wrong. (Rightly, he doesn't think that this will happen.) The CCR-generalization, that is to say, is not disturbed by the existence of irrationalistic doctrines, even of reinforced dogmatisms. The adherent of CCR, to be sure, is obliged to render such a position criticizable if he wishes to adopt it. If he is unable to do this, he had better drop it (without necessarily adopting its negation). In other words, CCR as a methodology is indeed in opposition to irrationalistic, non-argumentative methodology. But Bartley never said that a rationalist is able with integrity to adopt any attitude, any method. His own method he may criticize, explore, even come to condemn; but not in an irrational manner.

It will be accepted in d. below that there may indeed exist some positions that for logical reasons a rationalist cannot adopt, though tautologies and reinforced dogmatisms are not obviously among them. But, it will be argued, that does not seriously affect the CCR-generalization properly understood.

c.   Vacuity of CCR

It might be objected that the CCR-generalization is vacuously true, even analytically so. One way of reaching this conclusion, due to Watkins, need not detain us longer than it detained Watkins (*op. cit.*, 274). It is simply that since any hypothesis is contradicted by something (for instance, its negation), every hypothesis is potentially open to criticism. Well, yes, we might say, if a statement is allowed to count as a criticism even if it is known to be false, that is indeed so; but there is little in CCR that invites us to allow this. A better response is to observe, once again, that CCR is primarily a methodological doctrine, and that criticism itself has to be characterized methodologically. Statements, in short, are not criticisms; methods of attack are. We have seen this already in the case of logically true statements, which are in conflict only with logical falsehoods; nonetheless, they are open to potential (but inevitably unsuccessful) criticism by a method that applies just as efficiently to some contingent propositions.

It is on this account that empirical science is such a powerful critical tool. Through its requirement that physical effects be repeatable, science ensures that criticisms are not isolated statements of fact but the results of the application of general methods. In science we know how to test a statement of whose truth there is not the slightest genuine doubt. (Students do it all the time.) The same holds, though less spectacularly, in other areas too, even philosophy. That is what criticism is.

There is another argument that concludes that criticizability is trivially universal: Either a hypothesis is criticizable or it isn't. If it is, there is a potential criticism of it. If it isn't, that is an actual criticism of it. In either case there exists a potential criticism of it. Ergo, any hypothesis is criticizable. As Bartley himself observes (1984, 241f), this is not a good argument. It equivocates on the question of what standard we should apply in criticism. Bartley was the first to stress that the traditional way of criticizing a statement was to ask if it had been justified. CCR was meant to knock this idea on the head, by restricting all criticism (in the realm of fact) to the question of the truth of the statement under examination, not its epistemological (or even methodological) status. The fact that a hypothesis is not criticizable in no way affects the claim of the hypothesis to be true. Although, therefore, the seeming non-criticizability of a hypothesis may be a black mark against the procedures by which it is being investigated, it is not a criticism in the sense that CCR requires.

Think of the comparable case with regard to the demarcation of science from pseudoscience. We require the hypotheses admitted to science to be falsifiable—for otherwise we have no empirical way of getting rid of them; we are stuck with them, true or false, forever.

Accordingly we use a non-empirical method to determine what can be considered by the empirical method. Now an objector might urge that we should be prepared to admit any hypothesis we like into science, but that if its unfalsifiability is established—perhaps by mathematical means—we should push it out again, on the grounds that it is a scientific defect to be unfalsifiable. Obviously these two methods will be practically equivalent, except that the objector classes as an empirical activity what I would rather class as non-empirical; for it nowhere involves either observation, or experiment, or any recourse to nature. So with the more general case. There is no great harm done if we expand the domain of criticizability so that, quite trivially, all statements are criticizable. But CCR wants more: the truth of a statement has to be something that can be critically investigated. Without that, there is no comprehensively critical rationalism.

Thus I reject the thesis that criticizability is an automatic property of all statements. It is not an intrinsic property of statements at all, but an honour that must be bestowed on them by the development of appropriate methods of criticism. How, it may be asked, is this to be done? In many cases the answer can be only: by a consideration of the problems that provoked them (Popper 1963, 198–200; Bartley 1962, 159–161; 1984, 127) Those statements that are the response to no problem are accordingly the statements that it may be most difficult to criticize.

d. *Paradoxicality of CCR*

A more subtle objection is the claim, advanced by both Post and Watkins, that CCR is susceptible to something superficially similar to the paradox of the liar. The liar paradox has a structural similarity to Gödel's theorem of the incompleteness of elementary arithmetic, and Post claims to espy a parallel kind of incompleteness in CCR. Indeed one of his papers is entitled 'A Gödelian Theorem for Theories of Rationality' (1987; an earlier version is dated 1983). The crux of the matter, according to both Post and Watkins, is not that the CCR-generalization is not criticizable, but that the statement 'The CCR-generalization is criticizable' is not criticizable; in Watkins's account, this amounts to the claim that 'The CCR-generalization is criticizable' is analytic; and that the CCR-generalization is accordingly either analytic or false.

These arguments have gone through a bewildering number of

different variations, too many to be worth pursuing. What I shall do here is look at the version of the argument that Bartley sets out in Appendix 4 of his 1984 (and also in his 1987); an argument that, he says, "is thoroughly inspired by Post's, but perhaps simpler" (224), and uses no assumptions that he finds unacceptable (218). For this purpose the CCR-generalization is formulated as

> All positions are open to criticism.   (B1)

Now the point of CCR is that it comprehensive; it is a position that is open to criticism.

> B1 is open to criticism.   (B2)

Now if we were to show that B2 is false then we should have shown that B1, from which it follows, is false too. That is, we should have criticized B1. Since this possibility is what B2 envisages, B2 would be true. Every argument designed to show that B2 is false succeeds in establishing B2. "Any attempt to criticize B2 demonstrates B2; thus B2 is uncriticizable, and B1 is false" (Bartley 1984, 224; 1987, 320).

Post's argument proceeds along similar lines (Post, *op. cit.*, 254f). Instead of B1 and B2 he proposes the more involved

> Every rational, non-inferential statement is criticizable.   (P1)

> P1 itself is criticizable.   (P2)

(Here "non-inferential" means something like 'not part of the background inferential apparatus used in the development of criticism'. This plays no role in what follows.) According to Post, every criticism of P2 is a criticism of P1—because P1 implies P2—but no criticism of P1 is a criticism of P2—because a criticism of P1 confirms P2, perhaps even verifies it. (I ignore here some subtleties in Post's paper.) Every criticism of P2 is thus a non-criticism of P2; that is, there is no criticism of P2. It follows that P1 is false.

There is little to choose between these two arguments as arguments, although at first sight it might look as if Bartley has to assume that demonstrable propositions are uncriticizable (a view rejected in a. above). This is not so. All that is assumed is that demonstrating B2 is not a way of criticizing B2; and this is a component also of Post's argument. Nor is it implicit in Bartley's argument that "if a position is false then there is an argument showing it to be false" (Post *op. cit.*, 265, emphasis suppressed).

Bartley's surprising response to this kind of argument is to be unsurprised. He reminds us that CCR is formulated in natural language and asserts (not very accurately, as Post *op. cit.*, 265, remarks) that "Tarski has shown that any natural language that contains semantic terms and thus the possibility of generating self-reference may be expected to be inconsistent" (1984, 224; 1987, 321). If we wish, Bartley says, we may adopt any one of a number of methods to exclude the usual contradictions, and in this way undermine the whole framework propping up Post's argument. This is quite enough, he maintains, to re-admit the possibility that B2 and P2 can be criticized, without the consequential embarrassments that follow if Post's argument is valid. Post's argument is thus defeated.

Like Post *loc. cit.* I don't think that this response will do. For one thing, as is made plain by Post's comparison of his argument with Gödel's theorem (rather than with the liar paradox), 'criticizable' is not (or anyway, not obviously) a genuine semantic term; and so the appeal to Tarski's solution (or some other solution) to the semantical paradoxes is not to the point. But more generally, Bartley's response is too weak. No hint is given of how Post's argument and his own might be invalidated, or how B2 or P2 might be criticized. One might even say that Bartley's response is not properly a criticism of Post's argument, but a bare appeal to possibilities; the critical rationalist equivalent of the worst excesses of scepticism.

My own opinion, however, is that neither Post's argument nor Bartley's is valid; and, moreover, that Post's P1, unlike Bartley's B1, cannot be regarded as an even approximately adequate statement of the core of CCR. The trouble in each case is that the second displayed statement (B2, P2) is not, in my view, a consequence of the first (B1, P1). Before seeing why—the grounds are slightly different in the two cases—let us look at an even simpler argument, which cannot be invalidated on these grounds. The materials of the argument are the statements

$$\text{Every statement is criticizable} \qquad \text{(M1)}$$

$$\text{M1 itself is criticizable.} \qquad \text{(M2)}$$

By the same reasoning as before we may conclude that M2 is not criticizable, and that M1 is false. In other words, we cannot hope to express the CCR-generalization satisfactorily by M1, even if some loose formulations are little different from it. Post, as we have seen, weakens M1 to P1; Watkins weakens it to 'A comprehensively critical rationalist can hold open to criticism all the nonanalytic statements that he rationally accepts'. Bartley adopts B1 instead of M1.

What are the differences? I maintain that the crucial difference between Bartley's formulation of the CCR-generalization on the one hand and those of Post and Watkins on the other does not revolve around the explicit reference to rationality (implicit, I think, in the CCR-generalization when it is understood methodologically) but in the use of the word 'position' rather than the word 'statement'. CCR must not be understood to hold that every statement that a comprehensively critical rationalist counts as true (rationally accepts) is on its own criticizable. P1 must be rejected as a statement of the CCR-generalization (as must Watkins's variant). In fact I think that P1, like M1, though less transparently, is false, and rationally unacceptable; which means that P2 is not a consequence of it. In contrast, B1—which talks about positions, not about statements, is rationally acceptable; but B2 does not follow from B1 because B2 is not in the ordinary way of things what we call a position—it is just a statement. (Those who prefer to reserve the word 'statement' for what is called a 'speech-episode' would presumably deny even this. See Strawson 1950/1971, 190–93.) This may sound like sophistry, but really isn't. As far as statements (or utterances, or whatever they are best called) are concerned, what is important for the rationalist, I suggest, is that each statement that he accepts either is itself criticizable or follows from a statement that he accepts that is criticizable. Any position adopted must be criticizable, but it is no concession to the irrationalist to allow that some logical consequences of the position may not be criticizable.

To see that this manoeuvre is not just an ad hoc evasion, we return for a moment to the demarcation between science and metaphysics. As noted towards the end of 1.2, it has long been pointed out that science has metaphysical content. Does this show that the demarcation criterion is incorrect? Surely not. (But Bartley seems to think that it does. See his 1984, 203–05.) We must add a rider to the demarcation criterion as follows: unfalsifiable statements that enter science merely as the logical consequences of falsifiable ones must be rejected if these falsifiable statements are rejected. (The families of diplomats receive diplomatic immunity, not in their own right but by courtesy. If the diplomat loses his immunity and his right to stay in the country to which he was posted, so does his family, even though guilty of no offence.) A good example is metaphysical determinism, which entered science on the of Newtonian theory and should have been packed off on the same route. This is clearly in some sense a non-empirical mode of treatment, as noted in Chapter 1. But since we certainly want to count as true the consequences of any scientific hypothesis that we count as true, something like it is essential.

The same applies more generally. The comprehensively critical rationalist may not take up any position that he is not prepared to subject to critical examination. But though the critical examination of a hypothesis is normally conducted by critically examining some of its consequences, it is not conducted by examining them all. The rationalist is not at all obliged to try to criticize all the consequences of his ideas, and if, as may happen, some of them are not open to criticism at all—which may mean no more than that he cannot think of any way, even potentially, of criticizing them—that need not disturb him. This point could have been made explicitly by Bartley in 'The Transmissibility Assumption' (1984, Appendix 6), but inexplicably it was not.

In these terms B1 is a position that Bartley recommends that we adopt, and it is essential that it be criticizable. But B2 is just a consequence of it—an interesting consequence, in the light of what CCR says, and (one hopes) a true consequence; but it cannot be taken up as an independent position. Anyone who is tempted to adopt it needs only to look at the above arguments to see that, although B2 is a possible position on its own (though a strange one), it cannot be adopted at the same time as B1 is. Nor, of course, is there the slightest need for it to be adopted along with B1, which brings it along for nothing. All this is, no doubt, slightly unexpected, and one must be grateful to Post and to Watkins for bringing such curiosities to light. But they cannot be seen fairly as damaging criticisms of CCR.

e. The Uncriticizability of Logic

In addition to *a.* above, that logical truths are not criticizable, there is a much more serious point concerning the criticizability of our system of rules of logical inference. These are not, like tautologies, mere consequences of other positions we adopt, but vital constitutents of the rationalist position itself. Hence it does seem that CCR, though not perhaps committed to any doctrine, is at least committed to the use of logical rules of inference. Bartley's discussion of this problem in his 1984, Appendix 5, and elsewhere is confusing, I fear, and his claim that there are presuppositions of critical argument that are "hence not revisable in critical argument" (1984, 254; see also 135, or 1962, 173) is just the sort of thing that this critical rationalist does not want to hear.

As I see it, the situation is actually rather simple. Critical argument certainly cannot be carried on without some system of logic. You cannot in this sense abandon logic and remain a rationalist. Here Bartley and I are in full agreement. But the system of logic employed can—despite Lewis Carroll—be taken as one of the things under investigation in a critical discussion. I concede that if something untoward emerges we shall not normally blame the system that we are using. But sometimes we do—Russell did, for example, when faced with the logical paradoxes, and so did the intuitionists. The rules, after all, are universal and if they were systematically to lead to error in a way that in fact they do not, we might eventually decide that we had got them wrong. We must not be misled by the fact that we have not got them wrong. In any case, we know what a criticism in principle would look like, as I explained under *a.* above.

There is no reason why we should not use a computer language (say BASIC or FORTRAN) which is, essentially, a program run by the operating system of a certain piece of hardware, in a test of the correctness of the operating system. In an obvious sense the operating system is presupposed. But it does not follow that it is immune from criticism in such a test. "Logic is invincible because one must use logic to defeat logic" said Heaviside (quoted in Kline 1980, 316). Does anyone imagine that materials science is invincible because one must use materials science to defeat materials science?

If this is right, then the system of logic that we employ is itself open to criticism and modification. But, it may be said, that is utterly to miss the point of the objection, which is not bothered with the possibility that we may not yet have formulated the rules of valid inference in a universally correct way. The charge is that it is logic itself, not any particular formulation of logic, that CCR assumes to be beyond criticism. This line of objection thoroughly bemuses me. For even if there is such a thing as 'logic itself' (on which matter I am decently sceptical), logic itself enters into critical discussions only in some formulation (correct or incorrect); in no way is some unformulated system of 'logic itself' presupposed or supposed in rational inquiry. Bartley states that the key question is "whether we can revise logic in the sense of denying that true premises need always lead, in any valid inference, to true conclusions" (1962, 169; 1984, 132), and holds that the rationalist must answer it negatively if he wishes to remain a rationalist. It is of some interest that Etchemendy (1990) and McGee (1992) have recently suggested an affirmative answer in their criticisms of Tarski's theory of logical consequence.

Although his own responses to *a.–e.* were not the same as mine, I can believe that Bartley would have agreed with most of what I have so far said. (That is perhaps a barely criticizable statement, but it follows— given suitable other assumptions—from the fact that all that I have said is consistent with the main ideas of CCR, and none of it dislodges CCR's solution to the problem of providing a coherent solution to the problem of rationality.) But my last criticism is of a quite different kind. I am not confident that Bartley would have accepted it.

f. Reaction to Criticism

Though Bartley unendingly lauds the value and desirability of intense and sustained criticism, he tells us almost nothing about what to do with a hypothesis once it has been successfully criticized. At times (for example, 1984, 194–96), in his determination not to suggest that any criticism is final, he gives the impression that all we need to do is to take note of the criticism and pass on. This is especially obvious in his discussion of Popper's account of the empirical basis, where he writes (*op. cit.*, 215f):

> If such basic statements happen to be incompatible with a theory, then the theory is false *relative to them*; and they are false *relative to the theory*. There is no question of theory proving reports wrong. *Both* could be wrong: neither is 'basic'. . . . One contributes nothing to this situation by adding a requirement that one needs to decide by agreement which reports to accept.

This reluctance on Bartley's part to say firmly that criticized beliefs have to be rejected is doubtless the source of Musgrave's caricature of CCR as a doctrine according to which "a belief is rational if it is *criticisable*" (1991, 30; see also his 1993, 296). Musgrave notes that he has "probably . . . misunderstood CCR".

The problem, Bartley says, is to be traced to "Popper's unfortunate tendency to demand convention or irrational decision whenever some point is reached which cannot be justified" (215). But there is no question of such a decision's being irrational, and it is alarming to find the inventor of CCR suggesting any such thing. On the contrary, such a decision—which is always taken in the light of what we think is true—is itself open to criticism, emendation, and even reversal. Without such decisions, not just in the case of basic statements but throughout our knowledge, our criticisms are impotent, leaving their trace only in metalinguistic reports to the effect that this has been criticized by that. We need more if we are to accomplish (even fallibly) the task of separating truth from falsehood. But there is no danger that the decisions we make about what to accept and reject are arbitrary. They are constrained throughout by our desire to discover the truth, to separate it from falsehood. For all Bartley's interest in the truth, and his disdain for justification, he seems, by dubbing as irrational those unjustifiable decisions that we make in our quest for truth, to have lost his nerve at the very last minute.

5

Hume: Bacon = Gödel: Hilbert

5.1. Introduction

Hume was to Bacon what Gödel was to Hilbert. Startling as this comparison may initially seem, I shall try to suggest that there is much illumination to be gained from it; that by no means is it just a forced likening of unlike *tours de force*. But almost all the illumination is on one side—on inductivism, not on formalism. For whereas philosophers of mathematics—except one or two crackpots—have bravely taken the medicine dispensed by Gödel's demolition of Hilbertian formalism, and have been fortified, philosophers of science—except one or two crackpots, such as myself—have been less disposed to face up to Hume's demolition of Baconian inductivism. Most current theories of scientific rationality, I shall suggest, stand at about the same level of wistful thinking as fantasies that the problem of the consistency of arithmetic will in time yield to finitistically acceptable methods. I shall also outline, as incisively and as clearly as I can, what seems to me the only proper response to Hume and all other desperate sceptics and irrationalists: an abrogation of the search for justification, joined with a pitiless determination to exploit reason in the pursuit of truth. Only with such an approach can any decent progress beyond Hume be expected. I remember hearing Kreisel in characteristic vein (at the 11th International Wittgenstein Symposium at Kirchberg in 1986) showering scorn on those whose first (or only) reaction to Gödel's theorem is to gush. Philosophers need to gush less at Hume.

5.2. Hilbert and Gödel

The set-theoretical paradoxes associated with the names of Russell, Cantor, and Burali-Forti showed that mathematical theories may seem winningly plausible, intoxicatingly beautiful, and abundantly fertile, and yet be rendered uninhabitable by inconsistency. Cantor's paradise of infinite sets, as Hilbert described it, was one such theory: unfettered licence in set construction led to sets that could not be sets, to paradox, to a threatened devastation of the entire habitat. Planning laws became urgently needed, not only in set theory—where they were eventually provided by Zermelo and his successors—but, it was considered, in all other mathematical domains. One suggestion, that of intuitionism (more generally, constructivism), was that the methods of mathematics be limited to effective ones—to methods within our control. Intuitionism would have cleared away large parts of traditional mathematics. In contrast, Hilbert's own proposal—often called formalism, since its implementation called for the precise formalization of any theory that was to be scrutinized—was the indulgent one that any mathematical theory be permitted whose consistency could be demonstrated in a sufficiently meagre and secure metatheory. Mathematicians were to be granted freedom of activity as before, with free transport to infinite domains if wanted, but their compositions were to be screened and their excesses eliminated by an ascetic priesthood of finitist metamathematicians. That was Hilbert's plan.

As we know, the plan foundered. In 1931 Gödel showed that even the full resources of elementary arithmetic—which far exceed what Hilbert accounted as finitistically admissible—are insufficient for a proof of the consistency of elementary arithmetic (provided, of course, that elementary arithmetic is consistent). There is no real hope, therefore, for a finitist proof of the consistency of analysis, still less one of the consistency of set theory. The consequences of Gödel's work have been almost universally accepted: unquestionably secure proofs of the consistency of consistent mathematical theories are not to be anticipated. It is different with inconsistency: there will always exist an irreproachably finitist inconsistency proof for any inconsistent elementary theory. (We may not be long-headed or long-lived enough to uncover it.) Nobody draws the conclusion that because arithmetic and Zermelo/Fraenkel set theory, and other theories, cannot be glorified with finitist proofs of consistency, the question of their consistency is idle or without meaning.

Indeed, many consistency proofs using methods more adventurous than those Hilbert sanctioned have been given; familiar examples are Gentzen's proof of the consistency of arithmetic; Gödel's own proof using computable functionals of finite type (1958; for a helpful outline, see Troelstra 1990); and many relative consistency proofs in set theory. These proofs do not, to be sure, provide what Hilbert wanted—an unchallengeable certificate of freedom from contradiction. But they are tests of consistency nonetheless, and the methods they use, though not finitist, are not greatly different from standard mathematical methods. Indeed, it should be stressed that, even in the early days of mathematical logic, there was no presumption that all mathematical investigation of mathematical theories need be finitistic. In the Introduction to his dissertation—in which the semantical completeness of elementary logic is proved—Gödel writes: "this problem could have been meaningfully posed within . . . naive mathematics [that is, before the paradoxes appeared] (unlike, for example, the problem of consistency), which is why a restriction on the means of proof does not seem to be more pressing here than for any other mathematical problem" (1929, Section 1, 65). Dreben and van Heijenoort (1986, 46, 52) record that a similar view was held by Emil Post. In spite of this, there continued to be an interest in what could be done effectively or constructively. Logic, in particular, was expected to be effective.

5.3. Bacon and Hume

In *Novum Organum* (1620) Bacon characterized well-conducted empirical science as the advance by induction from "senses and particulars" to "the highest generalities" (1620, I, xxii) about "the inner and further recesses of nature" (*op. cit.*, I, xviii). Urbach (1987, Section 2[vi]), has disputed the standard attribution to Bacon of the doctrine that induction delivers incorrigible certainty. But it is not necessary to broach this issue here. If it was not Bacon it was someone else (Shakspere, perhaps); and what Watkins (1984, Section 4.2) calls the Bacon/Descartes ideal—it could be called the Parmenides/Aristotle ideal—certainly true theories of abundant explanatory power—is an aim of science that many adherents of induction have been inclined to favour. More to the point, Bacon did surmise that the inductive method ensured moral certainty, justification, or good reasons for the theories it throws up (Urbach, *loc. cit.*); and it is exactly this surmise that Hume (1739) refuted.

Traditional scepticism destroyed the view that our knowledge of the world could be certain or conclusive. An item of knowledge, in order to be certain, has to be proved to be true. But every proof (except a proof of a triviality) uses premisses; and though the premisses only need to be true for the conclusion to be true, the conclusion is proved to be true only when the premisses are themselves proved to be true—that is, themselves certain. (For more on this, see 3.3 above.) In modern terminology we say that although any statement may be the conclusion of a valid derivation, what is needed for certainty is a demonstration. Since a derivation may be converted into a demonstration only by demonstrating its premisses, something always remains to be demonstrated. No demonstration is complete. All this applies uncontroversially to the finitist proofs of consistency that Hilbert envisaged; secure they were hoped to be, but not infallible. The possibility of hallucination can never be wholly discounted even in mathematics.

Hume's distinctive contribution to scepticism about induction was to observe that even if all reports of sensory experience are classified as immediate knowledge, and on this score counted as certain (and therefore needing no demonstration), still no universal statement, or even singular prediction concerning the future, becomes certain; that is, no universal statement, and no prediction, can be validly derived from the class of all true reports of observation. No such derivation is valid.

5.4. Gödel and Hume

Gödel showed that there can be no finitist proof of the consistency of any mathematical theory. We may not inaccurately describe Hume as showing that there can be no finitist proof of the truth of any scientific theory. The term 'scientific theory' must here be understood to cover universal theories and singular predictions, and the term 'mathematical theory' to cover theories strong enough to include Peano arithmetic, or at least Robinson's Q. (In each case the theories may be described by the patter 'Principia Mathematica and related systems'.) I mean to look hard at this parallel between Hume's argument and Gödel's, with the intention of suggesting that the only colourable response to Hume is quite parallel to the standard response of logicians and philosophers of mathematics to Gödel. Nearly all philosophers of science, I shall contend, have recoiled from Hume's onslaught on Baconian inductivism in a manner that no mathematical logician would consider seriously as a way of proceeding in the aftermath of Gödel's second incompleteness theorem. This is not, I realize, much of an argument. I submit the comparison only as an illustration.

To be safe, I must emphasize that this analogy between Hume and Gödel does not depend in any way on the incorrect view that truth and consistency are properties of theories that are on an equal footing. I hope that it is sparklingly clear that in general they are not; and that it is because they differ that the analogy is only an analogy. Truth is a much stronger property than is consistency; and what is more important, it can be a material or contingent property whereas consistency is always a formal or necessary one. The falsity of a scientific theory, if false, and the inconsistency of a mathematical theory, if inconsistent, may both be established finitistically; but while in the first case the proof is material, an empirical falsification, in the second it is formal, the derivation of a contradiction. Still less does anything I assert here endorse Hilbert's opinion that truth in mathematics is reducible to consistency: "I may say that a . . . proposition is true as soon as we recognize that no contradiction results if it added as an axiom to the propositions previously found true" (1905/1967, 135, Principle I).

This has to be said, and stressed, because discussions of human knowledge habitually and perversely confuse truth and certainty, a material property and a formal property. Doubtless showing a statement to be true does not differ much from showing it to be certain; and the illocutionary force of uttering the statement in normal conditions may be to say that it is certain, or anyway known. This does not mean that truth is the same as certainty. It is especially painful to see truth and probability treated on equal terms, and, even more commonly, verisimilitude and probability. A typical example is this (O'Hear 1989, 120): "In other words, does not a theory which, through its explanatory account of reality, actually leads us to more empirical knowledge, gain some degree of confirmation or truth-likeness, by virtue of that very fact? The realist is arguing here that the success of the theory is to be explained in terms of its truth or probability, meaning that what it says about the unobservable world has a good chance of being true." (For countless interesting historical asides on the distinction between truthlikeness and probability, see Popper 1963, Addenda 6 and 7, 399–404, and also Niiniluoto 1987, Section 5.2, 160–64. For a comment on O'Hear's optimism here, see *i.* in 2.3 above.)

5.5.　Response to Gödel

The impact of Gödel's work was sharply and abruptly to redirect most work in the foundations of mathematics. No longer could it convincingly be held that the aim of research was to prove by finitist means the consistency of extant mathematical theories. This much was agreed even by constructivists, who were perhaps the least affected; for they had never believed this to be their task, having always been prepared to discard parts of classical mathematics that did not measure up to their exacting standards of proof. Most others, however, reconciled themselves to the fact that, though consistency continued to be as crucial, indeed indispensable, a property of mathematical theories as it ever had been, it was in many central cases not demonstrable by finitist methods. There seemed to be two activities that could replace the search for such absolute consistency proofs. One was to test for the consistency of a theory by examining searchingly its logical consequences—that is, to do mathematics. The other was to reduce the problem of the consistency of a theory to that of some other theory—again a typical mathematical adventure, though one that, it turns out, makes momentous and unavoidable use of that peculiar brand of mathematics ordained by Hilbert for his earlier task: mathematical logic.

　　Neither of these strategies offers any comfort to those who want their consistency really proved. An inconsistent theory—indeed there is only one—is capable of passing any number of tests of consistency; naive set theory in Cantor's hands, for instance, was prodigiously prolific but not monstrously so. On the other hand, relative consistency proofs have to assume the consistency of some theory near in strength to the one under investigation. To be sure, if a theory is proved consistent relative to some other theory then it must indeed be consistent. (Prove this!) But this tells us little about its consistency, for what we do not know, and cannot prove finitistically, is, for example, whether ZFC, Zermelo/Fraenkel set theory with the axiom of choice, is distinct from ZF. Cohen showed that their distinctness follows from ZF's consistency. But ZF's consistency remains unresolved.

　　One response to Gödel's work that I have never seen so much as mooted is that the aim of a finitistic proof of consistency be amended not by relaxing 'finitistic' but by relaxing 'proof'. The idea would be, I imagine, that we introduce some measure of the extent to which a finitistic argument, though not a proof of consistency, approximates to this ideal state. No such measure could easily be defined syntactically, since one erroneous step is all that any pseudo-proof can need. Perhaps it is feasible in some other way. I don't know. But however cleverly defined, such a measure would not advance Hilbert's programme one jot. For any theory that is almost proved consistent the question of whether it is consistent stays implacably open; and an incomplete proof (that it is consistent) would be no more readily deemed a proof here than it would be elsewhere in mathematics. Less inviting still is the proposal that a theory that has been almost proved consistent might qualify as 'almost consistent'. Heaven knows what independent meaning can be given to this expression, though there is some sense in describing a complete theory as 'almost inconsistent'. It may be that the systems of paraconsistent logic promulgated by da Costa and others (Priest et al. 1989) could be of some service here. But whatever 'almost consistent' might mean, a theory that is almost consistent but inconsistent is, from a classical viewpoint, simply and starkly inconsistent. A very radical revolution in mathematical practice—such as the replacement of classical logic by paraconsistent logic—would be needed before inconsistent theories could be tolerated.

It is apparent that in the wake of Gödel's theorem the aims of Hilbert's formalism were quietly, but comprehensively, dropped. The quest for consistency proofs was succeeded by a quest for consistent theories; proving gave way to probing. Finitistic investigation of mathematical theories ceased to promise secure foundations; but it enabled mathematicians to see what could not be proved, and became an integral tool of speculative research.

5.6. Four Responses to Hume

As noted, the aim of Baconian inductivism was to uncover certainly true universal statements about the natural world. The method was to be the systematic collation of data, amplified by inductive inference; the general truths sought were to be the conclusions of finitistic proofs from more or less unquestioned premises. (Whether the conclusions have to be the final elements psychologically as well as logically is not here important. I am quite content to agree that an empirical proof, like a proof in mathematics, will rarely emerge if the conclusion is unforeseen.) Philosophers sympathetic to the realism of Bacon's aim and to the empiricism of his method have reacted diversely to Hume's argument that, even granted the premisses (the evidence), no such finitistic proof is possible. But only one of these reactions, I submit, the falsificationism of Popper (1934/1959a) and his followers, much resembles the way in which the philosophy of mathematics was transformed after 1931. In this section and the next three I list twelve standard responses to Hume—the titles bestowed on a couple of them are somewhat graceless, I fear—, comment on them, and indicate their metamathematical analogues. A more general discussion takes place in 5.10. I begin with four responses that, though historically important, are seldom recommended today in the context of Hume's problem of induction. I shall hardly mention them hereafter.

a. Positivism

Strict positivists accept Hume's arguments, and conclude that the aim of science must be pulled back to what can be achieved by finitistic proofs. This austere doctrine, feigning no hypotheses, once embraced by Mach, perhaps, or Carnap, but presumably not now embraced for long by anyone, corresponds to constructivism and intuitionism in mathematics, but is vastly less interesting. Positivism deserts theoretical science, and in its effect is scarcely distinguishable from irrationalism.

b. Instrumentalism

This is the view that scientific theories, especially those involving unobservable entities, are, despite appearances, not really statements at all, but pragmatic instruments, to be judged only by the standards of applicability and efficiency. Hume's argument is considered not to have force; as scientific theories are not true, it is not disturbing that they cannot be proved to be true. Although instrumentalism has played some role in the history of mathematics—at one period it would have been eminently sensible to regard the theory of complex numbers as little more than a serviceable instrument—and even in metamathematics, it is not easy to transform it into an intelligible position when truth is replaced by consistency. It may be that the appropriate parallel here is provided by the Wittgensteinian whimsy that mathematics, like philosophy, is a repertoire of techniques, not of results; this meiosis is epitomized in such remarks as "[one] cannot calculate wrongly with a contradiction, because one simply cannot use it to calculate" (Monk 1990, 421; see also 438–42).

c. Conventionalism

This is the doctrine that the theories of empirical science (or for Poincaré 1903/1952, xxi–xxvii, 90–105, the most general of them, classical mechanics and Euclidean geometry) are disguised definitions; hence not only true, but even necessarily true—and superior to alternative definitions only because they are simpler. Hume's argument is side-stepped by draining the theories in question of empirical content, and thus in effect easing them out of empirical science. It does not seem that the virtues of this approach, whatever they may be, can be duplicated in metamathematics. An inconsistent theory cannot be made consistent, even trivially so, by adjusting the meanings of terms. Or rather, that can happen only if logic too is altered; but this is not what was at stake in classical conventionalism.

d. Apriorism

The Kantian doctrine of a universal a priori principle that in a valid way permits scientific laws to be inferred from empirical evidence abjures strict empiricism by relenting on the demand that proofs be finitistic. As such, it might seem comparable to the resort to non-finitistic methods in consistency proofs such as Gentzen's proof for arithmetic. (For a version of this proof, see the Appendix of Mendelson 1964.) But this comparison is grossly misleading. It is an immediate consequence of Gödel's theorem that Gentzen's transfinite method of proof, though strong enough for its appointed task, is not generally applicable; it does not suffice for all consistency proofs. In contrast, Kant's principle of universal causation is conceived as completely general, a sufficient validation of all inductive inferences. This is, no doubt, why no one to date has managed to formulate it, or any other principle of uniformity of nature, in a way that anyone else is seriously minded to take seriously (as noted above in 1.2).

5.7. Three More Responses to Hume

The next three responses to Hume relinquish in different ways the prospect of obtaining a complete proof of a scientific theory, but hold out hope that something incomplete might do.

e. *Inductivism*

Apart from the difficulty of stating a principle of uniformity of nature that is not glaringly false, the plainest objection to apriorism is that no useful principle could conceivably be a priori valid. Some authors, the earliest being Levi (1967, 3–6), have accordingly distinguished 'local' from 'global' justification and have proposed rules of local inductive inference or acceptance making no pretence to a priori validity, but nonetheless claimed (presumably) to be correct. In local justification, the evidence may contain previously accepted laws and theories, and is not restricted to reports of observation and experiment. This approach looks rather similar to the provision of relative consistency proofs in axiomatic set theory. Yet since inductive rules of acceptance are not deductively valid, the comparison is deceptive. Relative consistency proofs are valid derivations of one assertion of consistency from another. But applications of a local inductive rule are not pretended to be valid derivations of one assertion of truth from another. They could no doubt be made valid by reformulating the rule as a universal premiss. Unfortunately, such a reformulation would, it seems, take us back into global justification and the woes of apriorism.

f. *Probabilism*

According to probabilism, Hume was right that no scientific theory is made certain by any finite set of observational and experimental reports. But he was wrong, or those who so interpret him are wrong, to say that scientific theories cannot be made highly probable by empirical evidence. Induction is fallible, yes; after all, it is not deduction. But the extent of its fallibility can be more and more delicately controlled by the accumulation of more and more evidence. The aim of scientific activity, probabilism says, is to provide as much support or confirmation for our theories as we can: in effect, the assemblage of probable theories. (It cannot be said that probabilists have yet managed to show how the probability of a universal theory in any scientific language can reach even the subliminal height of $\frac{1}{2}$.) The corresponding orientation in metamathematics was dismissed in the penultimate paragraph of 5.5 above. A like shortcoming—that probable truth has nothing to do with truth—disposes of probabilism (as spelt out in detail in 1.1). Probabilists have, it seems, been enticed into thinking that, because *certain truth = truth attained with certainty*, it is also the case that *nearly certain (probable) truth = truth attained with near certainty (probability)*; and hence that the collation of probable truths has something to do with the collation of truths. It has not. A statement may be probably true but not be true, and the second equation fails. The search for probable theories, in contrast to the search for certain theories, is severed from, one might say orthogonal to, the search for truth.

g. Cognitivism

A superficially attractive variant of probabilism, which for want of a better name I call cognitivism, regards knowledge as the aim of science; that is, theories that, as well as being probable on (or perhaps well supported by) the evidence available, are also true. Cognitivism avoids the most dangerous objection to probabilism, that it denies to science an interest in what the world is like; rightly it says that science aims at the truth. Yet because this objection to probabilism is usually not noticed, cognitivism and probabilism are usually not differentiated. Indeed, cognitivism is standardly promoted (for example by Cohen 1980, Section II; E. Barnes 1990, 61f) as an improvement on the view that science aims at truth. But although the alleged virtues of knowledge over true conjecture have been puffed since Plato, what they are is confoundedly obscure (see 3.4 above). That knowledge (truth + probability, or even truth + corroboration) is a weird hybrid to espouse as the aim of science may perhaps be seen by contemplating its metamathematical analogue: consistent mathematical theories, almost proved consistent finitistically. (If a theory is consistent, who thrills to an unfinished proof of its consistency; and if it is not, is not an unfinished proof that it is consistent more of an insult than a consolation?) The mistake committed here by cognitivism is an artless anschluss of means and ends. Rigorous tests are constitutive of scientific method, but not of scientific achievement. A graceful style is perhaps constitutive of good golf; and adherence to the rules is constitutive of any sort of golf that is golf. But neither is a part of the aim of golf, which is just to go round in the lowest possible score. The bounder who scoops and lunges horribly but returns a score below par better achieves the aim of golf than does the elegant dilettante who never breaks 100. That is at least what the prize list says. In the case of some golfers, of course, style is valued as much as success, practice as much as progress, observance of precepts of method as much as attainment of results. Some investigators value proof as much as truth. My point is only that the importance of subjecting theories to stringent tests is easily explained by assuming that science aims at truth, and that there is no occasion for the cognitivist's insistence that support or corroboration be included amongst its aims.

5.8. Falsificationism and Some Variants

Of the reactions to Hume that hold fast to truth as an aim of science but radically reconsider the need for any brand of justification, the most important is falsificationism. It has been depicted in the large in Chapter 1 above, and is here given the kind of miniaturist-cum-impressionistic treatment that the other responses to Hume have had to endure. The two degenerate issue that follow are further discussed in Chapter 6 below.

h. Falsificationism

No secret is made here of the fact that falsificationism (more generally, critical rationalism) is much my favourite response to Hume's criticism of induction. What is proposed by Popper (1934/1959a; expanded on and developed in 1972, Chapter 1, and 1983, Part I, Section 2) is that, since no finitist proof of the truth of a scientific theory is possible, we do best to forgo proof as an objective, and devote our energies to collecting true theories. Truth being hard to recognize, there is a danger that our planned anthology of truths will contain unnumbered falsehoods by default. This danger is definitely not negligible. But it is definitely not neglected either. On the contrary, falsificationism requires that the most strenuous and honest exertions be made to identify, and to extirpate, error, by finitistic means wherever possible. (Sitting around complacently with a well-meant resolve to accept any refutations that happen to arise is a caricature of genuine falsificationism.) Falsificationism sees no percentage in trying to support, or justify, the theories that are not falsified. Justification is seen as a mission at once impossible, useless, and unnecessary (Chapter 3 above). The parallel stance in metamathematics was stated at the close of 5.5.

i. Possibilism

This, like cosmetic rationalism j., is the offspring of what I see as an unhappy union of critical rationalism and justificationism, differing

triflingly from critical rationalism methodologically—all three positions advocate an approach to science wherein it is an incessant interplay
of bold conjectures and ruthless refutations—but nonetheless betraying
dreadfully its leading ideas. According to Watkins 1984, Section 4.5,
truth is not a feasible aim for science, for we can have no finitistic proof
that it has been attained. The most that can be asked for as an
aim—Watkins labels it the optimum aim of science—is 'possible truth'
married to depth, unity, and content; 'possible truth' is what can be
attributed to theories that have survived merciless testing unscathed.
(Further discussion is postponed to 6.3 below.) This would be like
taking the aim of mathematics to be not consistent theories but theories
that have not been proved inconsistent. Possibilism reduces scientific
accomplishment to scientific method even more drastically than
cognitivism does. As I say, at a methodological level possibilism does
not differ very much from falsificationism. But what lurks behind it—a
demand that it be finitistically decidable which theory best achieves our
aim (Watkins *op. cit.*, 155f)—most surely does.

j. Cosmetic Rationalism

Musgrave (1989a, Section 3; 1989b, Section 3; and 1991) distinguishes
the business of justifying (or having a good reason for) a theory from
that of justifying (or having a good reason for) belief in a theory. As far
as the former goes, he agrees with critical rationalism: we cannot ever
justify our theories. But he holds that we can, and indeed must, justify
our beliefs in theories, utilizing for this undertaking some such principle
as this (1989a, 337): "it is reasonable to believe and act upon hypotheses
that have withstood our best efforts to show them false, even though
these best efforts are no reason for those hypotheses (or their consequences) and do not show the hypotheses to be true or even probably
true". To my mind, this is merely to pour stale justificationist wine into
new critical rationalist bottles (cf. Musgrave *op. cit.*, 316). The wine has
no spirit in it, for Musgrave does not divulge what use can be made of
such justifications and claims to reasonableness. Given that he offers no
independent meaning to the term 'reasonable to believe', it is hard not
to conclude that 'It is reasonable to believe *h*' is equivalent by definition
to '*h* has been savagely tested and has not succumbed'; which seems to
me merely to pay lip service to the doctrine that rationality requires
good reasons. But this interpretation cannot be sustained. Musgrave
thinks that his proposal (which he describes as an "epistemic inductive

principle") has teeth, as shown by his discussion of the question of how belief in it may be justified (*loc. cit.*; 1989b, 317–19). For a more penetrating discussion, see 6.4 below. As for the metamathematical parallel: if we have painstakingly studied a mathematical theory, and have not managed to reduce it to absurdity, then our failure plainly deserves to be reported. But to report as an additional item of news that we have found a good reason to believe in the theory's consistency seems to me to add nothing but words.

5.9. Two Final Responses to Hume

Some responses to Hume, such as *b.* and *f.*, claim that although science must give up aiming at truth, in some strange way it still aspires to theories that are justified or worthy of belief. Others, such as *h.* and *j.*, say that science must give up trying to justify its theories. It is possible to drop both truth and justified belief as aims; *i.* does, though it concedes some aim to science and, as *j.* does, advocates justificationist rationalism in theory appraisal. But apathy and indifference to how the world works can be taken two steps further.

k. Nihilism

According to nihilists, science has no aim, and must be characterized as what scientists, bound only by minimal rationality requirements such as consistency, actually do. The most widespread current version of nihilism is known as (personalistic) Bayesianism. Bayesianism is like probabilism in maintaining that scientists' (and others') degrees of belief are measured by probabilities; but, unlike probabilism, it sees no significance in very high or low probabilities. There is nothing objective, Bayesianism asserts, in the degrees of belief that agents acknowledge, beyond their conformity to the calculus of probability. A theory with a very high probability is no more noteworthy than one that is believed with moderate intensity. So much do some Bayesians desire to distance themselves from any discrimination among theories that is not founded on probability that they sometimes affect not even to understand what the word 'accept' means (Howson and Urbach 1989, 106, provides an example). The totality of scientific research, according to Bayesians, "the essence of the hypothetico-deductive method in general", is comprehended by the "transformation from prior to posterior probabilities" by

Bayes's theorem (the quotation is from da Costa and French 1989, 345, which is by no means a Bayesian pronunciamento). It is manifest that Hume's problem can have no impact on such extreme nihilism. I shall excoriate Bayesianism more thoroughly in 6.5. There is but one methodological principle that Bayesianism is prepared to endorse— Bayes's rule of maximizing expected utility. In Chapter 7 I shall argue how little sense this rule can make of science, of empirical investigation. As for the metamathematical analogue of this Bayesian or nihilist view of science, . . . it is left as an exercise. It is hardly worth writing it down, so glancingly does it impinge on any problem of metamathematics.

l. Naturalism

Sociologists of science, who choose to study science wholly naturalistically, seek to explain in terms of social or psychological laws even the constraints imposed by deductive logic or probability theory. The objective truth gives way to the adjective 'true', to what people call true. Along with all other problems of rationality, the problem of induction becomes an empirical issue, calling for a causal explanation. The last two sentences of the previous paragraph may be repeated, mutatis mutandis.

5.10. Conclusion

Gödel's work indicated with great force that the range of usefulness of finitistic metamathematical methods is much less than Hilbert had hoped. But it did not diminish the value of such methods when available. Their conclusions may be fallible, but we may conjecture that they are all correct. Luckily, as I noted above, any inconsistency proof is finitistic, being the derivation of a contradiction by effective rules. Any consistency claim for a mathematical theory is thus open to finitistic rebuttal. Similar things may be said about scientific theories; matters are less clean, however, since statements of empirical evidence are themselves vulnerable to empirical criticism. In addition, what are informally thought of as scientific theories must often be enriched with auxiliary hypotheses in order to become testable empirically. Nonetheless we are sometimes able to deduce an enlightening conclusion from the empirical evidence alone, and for this we can be grateful.

The critical rationalist or falsificationist is primarily concerned with criticism, in improving our knowledge rather than consolidating it, but he will welcome proofs where proofs are possible, and empirical verifications where they are possible. After all, they indicate directions in which, short of some more radical upheaval, improvements are not to be found. But he does not invest proofs, even the most elementarily finitistic of them, with any epistemological significance. As explained in 3.3 above, what we call proofs are all fallacious as proofs, and the job they do is not one of providing justification for their conclusions. It is in this light that we should evaluate one recent reaction to Gödel's second incompleteness theorem, to the effect that "[to] trust the consistency of a system on the grounds that it can prove its own consistency is as foolish as trusting a person's veracity on the grounds that he claims that he never lies" (Smullyan 1992, 109; see also his 1987, 111). Smullyan concludes that "the fact that . . . Peano arithmetic, if consistent, cannot prove its own consistency . . . does not constitute the slightest rational grounds for doubting the consistency of Peano arithmetic" (*loc. cit.*). We may disregard here Smullyan's disregard of the difference between a finitistic proof of consistency and a proof utilizing the full resources of Peano arithmetic, though there is a recognizable difference between trusting a speaker who says that he never lies and trusting one who says on oath that he never lies. But in other respects the conclusion is correct, as is its analogue in the case of empirical science. The invalidity of inductive arguments does not constitute the slightest rational grounds for doubting the truth of scientific theories. Nonetheless, many empirical theories are false, and many mathematical theories are inconsistent. The failure of proof to the contrary is not an excuse for complacency.

  It cannot be denied that the proportion enunciated in the title of this chapter is not a perfect one, and the chapter itself is accordingly less of an argument than a challenge to anyone who embraces some response or other to Hume's scepticism. Does a similar response to Gödel's second incompleteness theorem make any good sense; and if not, wherein lies the difference? (Devotees of Reichenbach's so-called pragmatic vindication of induction should perhaps ask themselves why this response seems quite inappropriate as a response to Gödel's second theorem.) It is a question that all those interested in human knowledge should be required to face. If metamathematicians, of all people, can waive the traditional demand for proof without lapsing into a melancholy despair of reason, it is disappointing if those who are supposed to explore the wider and more tangible world of experience continue to seek refuge in the shadow world of justified true belief.

6

Three Lost Labours of Deductivism

6.1. Deductivism

Deductivism, as I want to understand the term here, is the doctrine that there exists no inductive logic, and that all inferences that are ever made in science are deductive inferences. Several of the responses to Hume sketched in the previous chapter qualify as deductivist. For example, what I called positivism is deductivist, since it regards the content of science as exhausted by observational and perceptual reports and their deductive consequences. In contrast, most of those theories of science that are called hypothetico-deductivist are not strictly deductivist, since they lay on top of the truly scientific activities of hypothesizing and deductive testing also some process of inductive confirmation, so that—in some mysterious way—the best tested hypothesis at the end of the struggle with experience comes out as the best confirmed, dignified and glorified by the empirical engagement. On this account I have never liked to watch falsificationism, for all its entitlement, being paraded in the livery of the hypothetico-deductivist household. It would be better called hypothetico-destructivism, so as to stress the fact that the logical (as opposed to psychological) purpose of the deductive effort is always critical: we deduce consequences from what we know not in order to expand or to consolidate our knowledge, but to liquidate it. Falsificationism is unbothered by the standard justificationist criticisms of hypothetico-deductivism. I am equally unimpressed by the objection, constantly made against deductivism, that living scientific theories are usually ramshackle untidy affairs, with loose ends dangling at every point, and far removed from the neat axiomatized deductive systems of formal logic. This is doubtless true. But however scruffy a theory initially may be, it can always be investigated deductively; attempts can be made to determine what its consequences have to be, and what its assumptions cannot be, and these attempts may lead—if we are quite lucky—to gradual improvement, to the gradual emergence of something like a deductive structure. Deductivism, as I see it, is an article of scientific method, concerned with the task of improving and correcting our theories; it is not so much a thesis about the way our theories are formulated, but about the way that they should be attacked.

In this chapter I intend to look more deeply—and perhaps more sympathetically—at three of the deductivist responses to Hume that were summarily charged in the previous chapter: *i*. possibilism (Watkins); *j*. cosmetic rationalism (Musgrave); and *k*. nihilism (personalistic Bayesianism). In their three ways these three positions all vehemently oppose any recourse to inductive inference, and endeavour to embrace a pure deductivism. That is admirable, and I applaud it. But, as we saw in 1.3, the attachment to a spurious and invalid process of inductive inference is only one of the things wrong with the inductivist and empiricist tradition. Just as damnable is the adherence to justification-ism, to the view that we must be able to give good reasons for or to justify (if only to some degree) the hypotheses that we hold or prefer. The three positions I shall consider agree on this point too as far as science itself is concerned. Indeed both Watkins and Musgrave agree wholeheartedly with falsificationism that scientific theories, and even statements of empirical evidence, are always and for ever conjectural, and cannot be justified to the smallest extent. Bayesianism too, I shall maintain more controversially, gives up any real attempt to identify the rational agent with the person who justifies his scientific opinions. But at an epistemological level a limited justificationism remains an integral part of each of the positions *i.-k*. I say "limited justificationism" because none of the positions discussed is committed to any doctrine of the unassailability, or even justifiability, of statements of evidence, so that justification means only justification relative to the evidence. It was claimed in Chapter 3 that such justification is unable to serve any epistemological purpose. But this is not to say that the quest for it may not be epistemologically injurious.

Watkins thinks that preferences among competing hypotheses have to be justified deductively by—that is, be consequences of—the available evidence, and is thereby jostled into painting a distressingly etiolated picture of scientific ambition. A supporter of falsificationism *h*. in science, he succumbs to something approaching positivism *a*. in metascience. Musgrave likewise thinks that it is not scientific hypotheses (or statements of evidence) that require justification but our acceptance of them. To justify such acts of acceptance he inserts a very general (and clearly not itself justifiable) principle. Again falsificationism at one level gives way to something much less attractive in metascience; in this case, what looks like a form of apriorism *d*. Bayesianism is different; in common with similar nihilistic theories, it solves the problem by simply giving up all attempts to fashion preferences among scientific hypotheses. By eliminating all preferences one effortlessly ensures that all preferences are justified. It is easy to be so dazzled by the brilliance of this solution that one fails to see that it is incapable of providing any further enlightenment.

6.2. Preferences among Theories

Acknowledging the incontestability of Hume's argument that no judgement of the truth of a universal theory is deducible from any available empirical evidence, Popper reformulated the problem of induction along these lines: "Can a *preference*, with respect to truth or falsity, for some competing universal theories over others ever be justified by . . . 'empirical reasons'?" (1972, 8). To this question he gave the answer: "Yes; sometimes it can, if we are lucky. . . . since we are searching for a true theory, we shall prefer those whose falsity has not been established." I do not think that much justificationist weight should be put on the use of the expression 'justified by . . .' here. The immediate context makes plain that it means at most 'deduced from . . .'. For example, the passage above is immediately followed by the following remark: "the central issue of the logical problem of induction is the . . . truth or falsity . . . of universal laws *relative to some 'given' test statements*" (*loc. cit.*). In two later passages Popper clarifies his remarks further (1972/1979, 372 and 1983, 20; for the sake of simplicity I have suppressed here all qualifications concerned with verisimilitude):

> Whenever I say . . . that we can have good arguments for preferring one competing theory to another with *respect to our aim of finding the truth* . . . I give an appraisal of *the state of the discussion* of these theories, in the light of which [one of them] appears to be preferable to [the other].
>
> Critical reasons do not justify a theory, for the fact that one theory has so far withstood criticism better than another is no reason whatever for supposing that it is actually true. But although critical reasons can never justify a theory, they can be used to defend (but not to *justify*) our *preference* for it. . . . Such critical reasons do not of course prove that our preference is more than conjectural. . . .

So although our preference for one theory over another may be determined by the state of the discussion, it is not justified by it. The state of the discussion may be illusive. Certainly it is not deducible from the empirical evidence that of two theories in competition, one refuted, the other not, the unrefuted one has, as far as truth is concerned, any timeless edge over its rival. The evidence implies only that the refuted theory should not be preferred; for it is false (provided that the evidence is true), so cannot be better (with respect to capturing truth) than the unrefuted one is. What is not excluded is the possibility that the unrefuted theory is also false. Thus a stance of indifference between the theories is not ruled out. It may be observed in passing (though this is not the situation that Popper considers) that if neither theory is refuted then again the empirical evidence implies nothing firm about which of them, if either, is preferable. This last observation exhausts the critical commentary of the section entitled 'Popper's Attempt to Solve the Problem of Induction' on pages 4–6 of Howson and Urbach 1989; and despite its disreputable justificationist ancestry it is still being freely advertised as one of the "overwhelming objections" to falsificationism: "It is not an acceptable answer, within the falsificationist account, to say that we ever only seriously consider finitely many alternatives, for there is no purely falsificationist reason for restricting the discussion to these" (Howson, 1993, 4). What is wrong with limiting attention to serious contenders (apart from its not having been justified to be right) is not said.

6.3.   Scotching Scepticism

Science and Scepticism (Watkins 1984) is divided into two unequal parts: it would not be too inaccurate, though it would be rather misleading, to say that it is the first part that is concerned with scepticism, and the second with science. Part One might have been entitled *The King is dead!*, Part Two, *Long live the King!* More accurately, Part One sets out what Watkins calls Humean scepticism, makes it clear how formidable is the problem of overcoming it, and discusses in painstaking detail an impressive list (4f) of unimpressive strategies that have, time and again, been offered as satisfactory ways of answering Hume: apriorism, transcendentalism, nondeductivism, probabilism, phenomenalism, vindicationism, pragmatism, and naturalism are all cited. It will be clear that the list overlaps considerably the twelve-fold enumeration of responses to Hume in Chapter 5 above, but it needs to be stressed that Watkins's discussion is much more thorough than my own. The upshot is that there is only one strategy, conjecturalism, that survives. Part Two of the book is then devoted to an attempt (xi)

> to find an answer to Hume, but one that accepts the validity of, and is not vulnerable to, his central negative thesis; an answer that resorts to nothing illicit, fishy, or fuzzy; no postulates 'proved' by transcendental arguments, no theology in the guise of assumptions about the simplicity of nature or the preattunement of our minds to nature, no attempt to coax out of the probability calculus something it cannot give, and no relaxation of the deductivist idea of a valid inference to accommodate invalid inferences.

Watkins's aim, as he aptly puts it, is "to succeed where Descartes failed: to submit our knowledge of the external world to ordeal by scepticism and then, with the help of the little that survives, to explain how scientific rationality is still possible" (*loc. cit.*).

I am in perfect sympathy with this statement of intent of Watkins's. I also agree completely with him that the only way out of the mess is to dump the idea that knowledge, even empirical knowledge, is "ultimately . . . derived from perceptual experience" (3); we must acknowledge that knowledge is "conjectural and is only negatively controlled by experience" (4). But I have much less sympathy with Watkins's own, assuredly novel, variant of the conjecturalist strategy, worked out in Part Two of his 1984, a variant that sacrifices truth (or even truthlikeness) as an aim of science, and replaces it with the bloodless relic of possible truth. At the root of my disenchantment is the question that Watkins claims to have been the stimulus to his investigation: the question "Why is the best *corroborated* theory the *best* theory?" (xv). Posed in such general terms, this question plainly begs for a justificationist answer, an answer that justifies a preference for the best corroborated theory over its rivals; and as such it deserves little consideration from a conjecturalist. Undue uneasiness about it, together with a narrow reading of Popper's 1972 reformulation of the problem of induction (discussed in the previous section), seems to have induced Watkins to transform his project into one of establishing that it is a deductive consequence of the evidence (or of the state of the discussion), and in that weak sense justified by it, that the best corroborated theory is in every case the theory that best achieves the optimum aim of science (1984, Chapter 8, especially 279, 304–06). Since this is plainly not so if the aim of science is truth, he was thereby led to move science's goalposts, to dignify possible truth, rather than truth, as the aim of science. It seems to me that this is a serious capitulation, and one that brings no earthly benefits.

Watkins begins his presentation of what the optimum aim of science must be by laying down five adequacy requirements he hopes to be uncontroversial. The first four of these are coherence, feasibility, ability to "serve as a guide in choices between rival theories and hypotheses", and metaphysical impartiality. The fifth, not quite happily phrased, is that an aim of science should "involve the idea of truth" (124). Watkins then sketches what he imagines everyone would endorse as the ideal aim for science, if only it satisfied these adequacy requirements: termed the Bacon/Descartes ideal, it calls for the ultimate explanation of the way the whole world works, an explanation that is absolutely exact, equipped with predictive omnicompetence, and on top of all this, certainly true. In Watkins's words (129):

> all empirical phenomena rendered explainable or predictable by deduction from true descriptions of initial conditions in conjunction with universal principles that are certainly true, ultimate, unified, and exact.

But virtually none of the bits of the dream seems attainable on its own (so the aim is not feasible), let alone in party with other bits (so the aim is not coherent); the demand for ultimacy, moreover, is not metaphysically neutral. So Watkins waters down the Bacon/Descartes ideal to the call for "science to progress with explanatory theories that are ever . . . (B1) deeper, (B2) more unified, (B3) more predictively powerful, and (B4) more exact" (133) than their predecessors. The demand for certain truth is at the same time reduced to a demand that our theories become (A) more and more probable. But there is no suggestion from Watkins that this condition (A) is on a par with the four conditions (B1)–(B4). Following Popper, Watkins holds (133–36) that the five requirements are not together coherent; and in any event, (A) is not feasible. Watkins calls this position, that "uncertain hypotheses about the external world can[not] be established as more or less probably true", probability-scepticism (58). He concludes that "what has to give, if the Bacon-Descartes ideal is to be rendered both coherent and feasible, is demand (A)" (135).

The familiar demand (A) for greater probability has however to be replaced by something, for otherwise the fifth adequacy requirement would remain unmet. In Chapter 1 I suggested that the obvious thing to do here is to return to plain truth as the aim of science, since this always was what lay behind the exaggerated emphasis placed on certain truth. But this is by no means the route Watkins takes. He writes (155f):

> (A*) Science aspires after truth. The system of scientific hypotheses adopted by person X at any one time should be possibly true for him, in the sense that, despite his best endeavours, he has not found any inconsistencies in it or between it and the evidence available to him.

(A*) is the ingredient to be added to (B1)–(B4) to produce the optimum aim for science, called (B*). It is, Watkins tells us, "the strongest feasible core that can be retrieved from (A)" (155).

What are we to make of this? We are told that science aspires after truth (the same form of words is used also on page 280), which makes truth sound very much like an aim, but we are told also that possible truth is all that we can feasibly extract from the unlamented (A). The immediate conclusion is that truth is not a feasible aim; after all, truth is stronger than possible truth. But Watkins doesn't mean this at all; when he introduces the adequacy requirements, he says quite bluntly that "an aim is infeasible if we know that it cannot be fulfilled" (124), which surely debars truth from being an infeasible aim. For though perhaps we cannot know that as an aim it has been fulfilled, we do know that as an aim it can be fulfilled. No, what Watkins is really trying to say here is that truth as an aim infringes his third requirement that it must "serve as a guide in choices between rival theories". For he writes (125):

> Suppose it were proposed that science should progress with ever ϕ-er theories . . . towards the (perhaps unattainable) goal of the ϕ-est theory of all. Suppose, furthermore, that a clear meaning had been given to claims of the form 'Theory T_j is ϕ-er than theory T_i', but that there is no possibility of assessing the truth or falsity of such a claim. Then this aim would be feasible: it is possible that a sequence of theories . . . is progressing in the right direction. But while we might *hope* that it is, we might equally *fear* that it is not. Such an aim would fail [the third requirement]. . . .

But it seems quite clearly wrong to say that "there is no possibility of assessing the truth or falsity of" the claim that one theory is true, another false, unless by 'assess' one means 'justifiably assess'. Surprisingly it seems that Watkins, who calls himself a conjecturalist, does mean this. For his central purpose, it turns out, is to defeat a form of scepticism that he sees as far more serious that probability-scepticism. By rationality-scepticism Watkins means the doctrine that "we never have any good cognitive reason to adopt a hypothesis about the external world" (58). In effect it is this doctrine that *Science and Scepticism* is out to counter. And the only way that Watkins finds to counter it is so to doctor the aim of science that we can *"know* [for Watkins the word 'know' in italics is a success word; see page 11], in suitable cases, which of the competing theories before us best fulfils that aim". Now because of the fallibility of the empirical basis (see c. in 2.2 above) we cannot *know* that any given hypothesis about the external world is true or false; nor can we *know* even that the hypothesis is consistent with the evidence (155); all we can *know* is that we have tested it thoroughly and have not found any inconsistencies between it and the evidence. So that is as much of (A) as we may include within (A*) if we wish to avoid the horrors of rationality-scepticism. The aim of science reduces to finding ever deeper theories that survive, as far as we can judge, our very best efforts at experimental and observational refutation. This is a feasible aim, but it is designed to be more than that. In the right circumstances we can be justified in asserting that one theory in our possession better satisfies the aim than does another.

It seems tolerably clear what has happened here, what has led Watkins to promote to the status of an aim what is in reality only a (fallible) means of attaining that aim and (to make room for it) to kick the real aim upstairs with the rank of aspiration. An intense aversion to inductive inferences has induced him to conclude that if we are ever to be able rationally to prefer one theory to another, if, that is, we are to repel rationality-scepticism, we shall have to do it with the machinery of deductive inferences only. So it will have to be possible to determine from the state of the critical discussion alone which of the theories on the table best fulfils the aim of science. A "good cognitive reason to adopt a hypothesis about the external world" (the kind of thing that rationality-scepticism denies the existence of) will be no reason at all for Watkins unless it is a conclusive or sufficient reason; otherwise it will be based on an inductive inference (or something fishy like that). And if there is no good reason for our preference, it is, according to Watkins, "a matter of uncontrolled guesswork" (287).

Now I share Watkins's dislike of counterfeit inference tickets. I share too a dislike of "uncontrolled guesswork". But I am an admirer of *controlled* guesswork. Indeed, I much prefer controlled guesses to good reasons. And I wonder why Watkins, who, no less than I, esteems bold and imaginative, but nonetheless controlled, guesses within science, is not prepared to countenance them when it comes to talking *about* science. Watkins quotes (*op. cit.*, 60) Lewis's dictum that "[if] anything is to be probable then something must be certain" (1946, 186) without passing judgement, but it seems pretty clear that there is a similar maxim lurking in the background of his own work: a maxim to the effect that if anything is to be a rationally controlled guess, then it must be controlled by something that is not a guess. I hasten to stress that Watkins does not endorse any such formula *within* science itself; the conjecturalist response to Humean scepticism, after all, undertakes to show how scientific knowledge is possible even though all synthetic statements are guesses. But guesses have to be controlled, and Watkins takes this to be a task to be undertaken at another level; since guesses cannot control guesses, we must control them with tough metascientific prophylaxes. Here is what he has to say about the problem of appraising judgements of comparative verisimilitude (287):

> But if all verisimilitude appraisals are guesses that are never positively justified, there is no more possibility of justifying . . . [one] appraisal than . . . [the opposite one]; a 'criticism' of the former conjecture would be no more than a rival and equally unjustified guess, one metaphysical claim pitted against another. . . . if verisimilitude appraisals can only be conjectural, and if two theories are comparable for verisimilitude, then a comparative verisimilitude appraisal of them cannot be under any genuine critical control.

But all this overlooks the possibility that guesses can be controlled, though not coerced, by other guesses. The genuine, if modest, success of liberal democracy shows that the theory of checks and balances can be made to work in politics. And it can be made to work in science too, if we care enough about finding out what the world is like; care enough about truth, that is to say. Popper (1945b, Chapter 7, and 1960/1963, Section xv) has shown how much political theory has to learn from methodology. There are important lessons in the other direction as well.

In opposition to Watkins's unnecessary slide back into justificationism I suggest that rationality-scepticism is a doctrine well worth embracing: indeed, "we never have any good cognitive reason to adopt a hypothesis about the external world" (Watkins *op. cit.*, 58). But scepticism is false. We do have ways, irreparably fallible ways of course, by which to separate true and false hypotheses. Irrationalism is false too. We may never have reasons, but we do have reason, and we use it all the time in the search for truth. All this is elaborated in Chapter 3 above. Moreover, rationality-scepticism does not imply probability-scepticism. Watkins asserts that "we would surely be out of the sceptical wood if we could, generally, select from a number of alternative hypotheses the one that is most probably true, given our present evidence" (42; see also 58). I can't see it. Indeed I don't see probability-scepticism as a version of scepticism at all. If truth is our aim, what probabilities our various hypotheses have is a matter of no significance.

What then of corroboration? "Why do corroborations matter?", according to Watkins (*op. cit.*, 279), is a question often posed to Popper by his critics but never answered in a satisfactory way; and it was in order to yield the answer that the best corroborated theory is the one that best fulfils the aim of science that the aim of science was trimmed down to (B*). Can those of us who reject the possibilist truth-component (A*) of (B*) give any better answer? It must be remembered that it is part of falsificationist epistemology that the passing of a test does nothing to secure, confirm, or in any way brighten the prospects of a theory. Nonetheless, the question can be answered satisfactorily.

The answer is that corroboration doesn't matter, even in the practical realm. It has no epistemological significance at all, as Popper always insisted. But testing matters, and has undeniable methodological significance. We want true theories. Testing is important because it is only by subjecting our theories to tests that we have any opportunity of eliminating those that are false; and the more severe the tests, the more generous the opportunity. We might put things this way: when a theory fails a test, we learn something but end up knowing nothing (since what we knew, our theory, has been eliminated). But when a theory passes a test (when, that is to say, it is corroborated), we learn nothing (since we already knew what the result of the test was going to be) but we continue to know something. Corroboration is doubtless needed if science is to exist, for if no theory were ever corroborated there would be no science; but it makes no contribution to the growth, or to the progress, of science. Popper's famous third requirement has to be seen as embodying not "a whiff of verificationism" (1963, 248, note 31) but what might be called a whiff of verisimilitudinism.

6.4. Cosmetic Rationalism

It was the thesis of 3.4 above that even if good reasons, or favourable reasons, were to exist, they would be useless epistemological embellishments, unable to assist in the classification of scientific hypotheses as true and false, incapable of adding to our practical decision making even a flicker of rationality not already earned without them. Nowhere is this redundancy of good reasons or of justification more glaringly obvious than in the version of critical rationalism expounded by Musgrave (1989a, Section 3; 1989b, Section 3; and 1991). In a recent introductory book he has called it fallibilism (1993, Chapter 15). This is regrettable, as fallibilism is normally understood to be the epistemological thesis that nothing can be known for certain; at best a preliminary to critical rationalism or falsificationism, it is in no sense a methodological thesis about how inquiry should be conducted. Like Watkins, Musgrave is a falsificationist at the level of science itself; he rejects all suggestions that scientific theories need to be, or can be, justified, whether conclusively or inconclusively; he rejects all modes of inductive argument (1989, Section I); for him scientific method is just the method of bold conjectures tempered by unstinting criticism (1993, 281). But like Watkins, Musgrave thinks that our preferences between theories, or our decisions about which theories should be accepted or believed, do stand in need of justification if we are to escape promiscuous irrationalism. He distinguishes sharply between justifying a theory (or belief) and justifying accepting it (or believing it); it may, Musgrave maintains, be justified (or reasonable) to accept a theory that is itself completely unjustified. A theory that succumbs to criticism must of course be rejected; but a theory that has withstood criticism may, according to Musgrave, be reasonably accepted or believed.

How can such reasonable or justified acceptance of a theory be possible if—as I maintained in 3.3—all justification that is not simply authoritarian is circular or question-begging? As I have noted in the previous chapter, Musgrave urges the rationalist to cleave to what he calls an epistemic inductive principle, a principle that baldly asserts the reasonableness of accepting well-tested and unrefuted theories; more generally, theories that have survived severe critical scrutiny. Indeed, he goes so far as to insist that "any deductivist who is not a *total irrationalist* must accept some inductive [but epistemic, not metaphysical] principle" (1989a, 332; the rhetorical emphasis is Musgrave's). With such a principle—let us name it *P*, though it is not the same as its namesake in 1.1—we may devise a facile little derivation of this form (1991, 26):

> *Epistemic inductive principle P*
> *H* has survived severe critical scrutiny.
>
> ――――――――――――――――――――――――――
>
> It is reasonable to accept *H*.

If, therefore, *H* has satisfactorily passed the tests to which it has been subjected, we shall, given *P*, be able to conclude that it is reasonable to accept *H*. The argument is unassailably deductive. The inductive principle *P* that constitutes the major premiss of course remains unjustified. But according to Musgrave this does not matter. What we need is not a justification of *P*, a reason in favour of *P*, which "would immediately require a reason for the reason and so on *ad infinitum*" (*op. cit.*, 29), but a justification for adopting *P*, a reason to accept *P*. And this, he suggests, is available, since *P* has been critically examined, and has survived the examination. By *P*'s own standards, therefore, *P* may be reasonably accepted.

Musgrave does not attempt to hide the fact that there is some kind of circularity or infinite regress at the heart of this. Conscious of the claims of Bartley's CCR (Chapter 4 above), he writes (*op. cit.*, 29):

> At this level of abstraction circularity is hard to avoid (any general epistemic principle will be subject to it) and the alternatives to it are even worse. One alternative is simply to postulate *P* and give no reason for adopting it. . . . The other alternative is to give a reason for adopting *P* . . . which is not of a critical rationalist kind. . . . On either alternative, the critical rationalist is not being a critical rationalist *about critical rationalism itself.*

Later he says that "to show that fallibilism is reasonably believed by fallibilist standards . . . is to argue in a circle" (1993, 294). But this is to let his position off far too lightly. The true situation of his principle *P* is much more unpleasant, as we shall see if we look more carefully at how it is supposed to be deployed in general, and how it may be applied to itself in particular. The trouble, very briefly, is that in his attempt to justify the acceptance of *P* Musgrave assumes the truth of *P*, though in reality he is entitled to much, much less. In the next paragraph, which is a little involved, I shall need to call also on a second principle that Musgrave enunciates (1991, 26), which I shall not combat: the principle that deductive consequences of reasonably accepted statements are themselves reasonably accepted. We may refer to this as the principle of transmission of reasonable acceptance.

We are given a report e that a scientific theory H has been well tested and has survived the testing unscathed. Let us allow that this evidence report e is not only true, but even known to be true. Notoriously there exists no deductive inference from e as premiss to H as conclusion. If we add P as a premiss, we may deductively infer that it is reasonable to accept H (though we may not infer that H is true). Now the validity of the derivation from e and P to 'It is reasonable to accept H' means only that if e is true (as we have allowed) and P is true, then 'It is reasonable to accept H' is true. To be in a position to conclude that it is reasonable to accept H, we would still need to know that P is true. But Musgrave admits that we do not know this, and that we cannot even have a reason for it (since it "would immediately require a reason for the reason and so on *ad infinitum*"); the most we have, he admits, is a reason to accept P. Using the principle of transmission of reasonable acceptance, we can extract from the valid derivation

Epistemic inductive principle P
e

It is reasonable to accept H

the conclusion that it is reasonable to accept that it is reasonable to accept H. But we do not seem to be able to extract any more. If this is right, then the most we can extract from any derivation of this type with P as major premiss is a conclusion with a doubly iterated modality: 'It is reasonable to accept that it is reasonable to accept . . . '. But then the same applies to the derivation

Epistemic inductive principle P
P has survived severe critical scrutiny.

It is reasonable to accept P.

Since we do not know the major premiss P, we may not use this derivation in order to conclude its conclusion, that it is reasonable to accept P. The most that we may conclude (assuming that it is reasonable to accept the major premiss) is that it is reasonable to accept that it is reasonable to accept *P*. In other words, Musgrave's epistemic inductive principle *P* can never be used to give a reason for its own adoption. The argument is not circular, but—as long as Musgrave's crucial distinction between a reason for *H* and a reason for accepting *H* is maintained—strictly regressive. The status of what can be extracted from the argument is necessarily more hedged than the status of the major premiss. To be sure, some subsidiary premiss could be trundled in at this point to permit *AH* to be derived from *AAH*, where *A* henceforth abbreviates 'It is reasonable to accept'. But the same argument that we have just waded through could be applied to this new premiss. It is presumably not analytically or logically true, and the best that we can hope for on its behalf is that there is a reason to accept it. This might make it reasonable to accept any conclusion of the form 'It is reasonable to accept *H*' obtained with its help, but it would not make it reasonable to accept *H*. In particular, it would not make it reasonable to accept *P*. And if it is not reasonable to accept *P*, then the schematic derivation presented above will not enable us to reach any interesting conclusion. (The substance of this paragraph is repeated in a more formal manner in the technical note at the end of the chapter.)

No reason has been revealed by Musgrave for the adoption of the epistemic inductive principle *P* (or any similar principle). Musgrave's justificationist version of rationalism, like all its congeners, is irrational by its own lights. But let us imagine this objection waived, and that *P* is not only rationally acceptable but actually accepted. (Perhaps someone will succeed in convincing us that the conditional 'If *AAH* then *AH*' is a logical truth.) Then we may be able to conclude of suitable *H* that it is reasonable to accept *H*. My question now is what we are to do with this conclusion. We are patently not mandated to accept *H*, or classify it as true; we are not even provided with a licence to infer it. Both of these would amount to the resurgence of an inductive rule yielding universal conclusions from particular premisses. There is a gap between the acceptability of the theory *H* and the theory *H* itself that Musgrave's variety of critical rationalism is powerless to span. (Compare what happens in Carroll 1895.) To be sure, there is nothing that prevents us from accepting *H* in a free act (or conjecture); and it may be that that is all that Musgrave intends, the conclusion 'It is reasonable to accept *H*' being no more than a commentary on what we are doing. But if this is so, it is clear that the claims to reasonableness, the good reasons, the justifications, are all window dressing. What we are really doing is indulging in free conjectures, which we continue to submit to severe testing. Whether there is any reason to conjecture H seems to be something that makes no difference to anything else.

I should emphasize that none of the above line of argument applies to CCR, Bartley's comprehensively critical rationalism (discussed in Chapter 4 above). In CCR no justification is deemed necessary for the acceptance of any position. We may conjecture whatever we like, provided that we then do our best to eliminate it. Musgrave briefly compares his position with CCR (*op. cit.*, 30), but he fails to bring out satisfactorily the real contrast between the justificationist weaknesses of his own point of view and the critical strengths of Bartley's.

In 3.3 above, I claimed that there exists no such thing as a good reason, conclusive or inconclusive, for any theory. Musgrave concurs, but despairs of rationality without good (inconclusive) reasons for accepting a theory. Though Musgrave is a staunch realist (as his 1988 shows), there is more than a glimmer of similarity between this position and anti-realism, where warranted assertibility usurps the throne of unwarranted truth. But it is truly a matter of no importance whether there are reasons for a theory, or for accepting a theory. Whatever they are for, whatever are they for? The rank inutility of inconclusive reasons is, I am afraid, not enough appreciated (though I have laboured it in 3.4). For even if we have a good reason to accept a theory, the inescapable question remains of whether it is true (or, if you like, whether to accept it as true). That is what we want to know, and no amount of justificationist flummery gets us any closer to what we want. It is difficult sometimes not to conclude that in Musgrave's hands the expression 'It is reasonable to accept H' is no more than a synonym for 'H has survived severe critical scrutiny' (so that principle P is true by definition). Musgrave denies this (*op. cit.*, 27), of course, but if there is a difference, it is a difference that makes no difference.

6.5. Pure Bayesianism

Criticisms of Bayesianism abound, and it may be wondered whether it is necessary to add to them. Surely every one of the 46,656 varieties of Bayesianism catalogued by Good in his 1971/1983 is vulnerable to at least one of the deadly objections raised in recent years. That may be so. My purpose here, however, is to focus on a single complaint whose force seems to me to be much too little appreciated. In a sentence, the complaint is that Bayesianism provides a solution to the problem of induction only by wholly abandoning interest in the battle for truth, and opting for a passivist theory of human knowledge that may roughly describe, but certainly does not explain, what happens when science is done. This abdication is rarely given the prominence it deserves. It is not even alluded to, for example, in the valedictory chapter of Howson and Urbach 1989, which sets out to answer the charge that "the subjective Bayesian theory . . . has been the object of much critical attention . . . [and] is still regarded in some influential quarters as vitiated by hopeless difficulties" (257). Nor does the massive failure of Bayesianism to make sense of science seem to be recognized by several distinguished critics of Bayesianism, who descry some of its shortcomings but are astonishingly unprepared to acknowledge how unsatisfactory a position it is. Earman, for example, says that he is a Bayesian on three days each week, and a non-Bayesian on three other days, and reports that "the upshot of my examination of Bayesian confirmation theory is neither a simple thumbs up nor a simple thumbs down" (1992, 5). Kitcher makes it clear that Bayesianism is not for him, but nonetheless thinks that "there are at least two ways in which Bayesian conceptions prove valuable" (1993, 293). Likewise, Glymour winds up his confession of why he is not a Bayesian with the words: "None of these arguments is decisive against the Bayesian scheme of things, nor should they be, for in important respects that scheme is undoubtedly correct" (1980, 93). I fully disagree; and it rather served Glymour right, I am afraid, when Rosenkrantz jumped up with the riposte that "Glymour . . . is a Bayesian, more so than many who march under that banner" (1983, 69). My thesis is that there is nothing whatever of value in the Bayesian position—which is not at all the same as saying that Bayesians have not often made valuable contributions to our understanding of science. The explanation for this profound disagreement concerning what Bayesianism accomplishes is quite simple: to my mind (but not apparently to the minds of others) Bayesianism is a philosophy that persistently strains at imaginary problems (especially 'the problem of scientific inference' and 'the problem of confirmation'). It is not so much wrong—Bayesianism is wrong—as wrongheaded. It leaves unexplained things that it ought to have explained, and explains things that it ought not to have explained. There is no health in it. Bayesianism does not explain the purpose of scientific activity, I shall maintain; neither theoretical nor empirical activity. Nor does it explain the objectivity and rationality of science.

Impurely stated, the pure variety of Bayesianism that I shall limit my attention to is the doctrine that, since the truth values of scientific hypotheses cannot be talked about with certainty, they had best not be talked about at all; instead, we should take notice only of the extent to which hypotheses approximate certainty, and we should take the probability $P(h|e)$ of a hypothesis h, in the light of the available evidence e, to register our attitude towards it. This does not mean that the more probable of two hypotheses is the better one, only that it is the one to which we apportion more confidence. There are a number of arguments that undertake to show that degrees of certainty should obey some more or less standard axioms of the calculus of probability, and I shall not contest them. The most important of the constraints that the axioms put on the probabilities is Bayes's theorem, a trivial result that gives Bayesianism its name. In Bayesianism Bayes's theorem is interpreted as a recipe for updating the probability of a hypothesis h in the face of new evidence e: $P(h|e)$, the new probability, should be equated with $P(h)P(e|h)/P(e)$. This is usually called (Bayesian) conditionalization. It is well known that the probability axioms themselves do not determine values for the function P except in some extreme and barely contentious cases, and how the probabilities themselves are to be evaluated is an issue over which Bayesians themselves differ. A minority opinion holds that rules can be laid down that determine in a rigorously objective manner the correct value of $P(h|e)$ for many h and e. (See Rosenkrantz 1977, Chapter 3.5, for a sharp statement of this position, and Seidenfeld 1979 for a sharp criticism of it.) Most Bayesians these days, however, are subjectivists or personalists, and hold that there is rarely one correct distribution for P; each agent is free to adopt any values he chooses, provided only that these values together obey the axioms. What both parties agree about is the need to be able to determine without risk of serious error what the value of $P(h|e)$ is: either by resort to objective rules, or by some operational procedure such as the juggling of bets. The above brief remarks, I stress, are meant only as a very superficial guide to the Bayesian philosophy, which enjoys an unusual degree of polymorphism amongst its various practitioners. Almost all the principles I have stated here have been modified or qualified, or even abandoned, by one Bayesian or another. But there remains a core position: forget about truth and falsity and put your trust only in the evaluation of probabilities.

I have explained at length in 1.1 why this is, for both theoretical science and practical action, absolutely the wrong retreat from the untenable doctrine that science pursues certainty. The correct move—I can hardly think of it as a retreat—is to retain the view that science pursues truth, and to acknowledge that the pursuit is itself uncertain: not only our hypotheses, but our evaluations of them, risk error at every turn. Bayesianism, in contrast, like the philosophy criticized in 6.3 above, hopes to palliate the unavoidable susceptibility to error at the level of hypotheses with some kind of watertight apparatus at the level of their evaluation. Unfortunately, as in the philosophy criticized in 6.4 above, the evaluations that can be made are irreparably detached from our original concerns. Certain truth deductively implies truth; but this implication dissolves when certainty is diluted to probability. Certain truth may be truth attained with certainty, but—because what is probably true need not be true—probable truth is not truth attained with probability. The connection between truth and probable truth is no firmer than that between truth and rumoured truth, or between truth and *ex cathedra* truth; or, indeed, between truth and improbable truth. That is to say, there is no logical connection at all. The best Bayesians recognize all this, and being good deductivists do not pretend to espy a logical connection where there is no deductive connection. Wishing to be alone with their beliefs, they therefore find that they must divorce themselves from truth. And so they do. Whether a given hypothesis is true or false is no longer a matter of concern to a pure Bayesian. Discovery of the truth is no longer the aim of science. This desertion from truth might be tolerable if some related aim (such as explanation, or the solving of problems) were put in the place of truth. But no such alternative is proposed. This failure to bear on the aim of scientific investigation is the main objection to the Bayesian position.

Properly speaking, of course, Bayesians do not have opinions or beliefs, only degrees of belief. What pure Bayesians call "learning from experience . . . is just the process of revising probability assignments in the light of additional information" (Rosenkrantz *op. cit.*, 48). The key to this process of revision is conditionalization, tiresomely called inductive inference by many Bayesians. But there is nothing at all inductive about Bayesian conditionalization. Statements of probability are not statements about the external world, and how they are amended in the light of new evidence is determined perfectly deductively; not as a deduction from the evidence, to be sure, but as a deduction from the current state of the evidence (in this respect Bayesianism and the theory criticized in 6.3 above are similar). Hume's problem of induction is solved in a minimalist way: since Bayesians never claim that a scientific hypothesis (apart perhaps from a statement of evidence) is true or false, they never have to justify such a claim. According to Jeffrey "there is no Bayesian or probabilistic theory of theoretical preference . . ." (1985, 141). Nor need there be, since no scientific theory is ever preferred to any other by the pure Bayesian. Of course, one theory may be more probable than another. But that is the end of it. Being probable doesn't make it a better theory (though what non-Bayesians call being a better theory might conceivably help to make it more probable).

There is doubtless something noble, and also a little spooky and unreal, about this rigorous renunciation of interest in what the world is like, this turning inward from the grubby uncertain world of guesses and errors to an aloof preoccupation with how one's degrees of belief about the world are modified by experience. Few Bayesians live up to the challenge of consistent asceticism. Even de Finetti, for example, guessed (without any probabilistic hedge) that "we shall all be Bayesian by 2020" (1970/1974, 2; the formulation of the prediction is from Lindley's foreword on page ix). Howson and Urbach say at one point that "no prior probability distribution expresses merely the available factual data" (*op. cit.*, 289), as though any distribution expressed any factual proposition. (Of course, one can read much into and out of a distribution, just as one can read much into and out of tea leaves.) The pure Bayesian should be prepared to acknowledge that personal distributions of degrees of belief just are what they are; rational to the extent that they conform to the probability calculus (including Bayes's theorem for updating), and no further; and unconnected with factual judgments about the world.

On this view, science has no aim. The acquisition of new information in general leads to changes in probability distributions, but it does not lead to improvements in them. Discovering an item of evidence that makes a hypothesis more (or less) probable is not a scientific advance; it is simply a move. Once it is maintained in addition that the search for confirming or disconfirming evidence is what is characteristic of science, all the austere attractions of the Bayesian position wither away. It is back there in the morass of inductivism with all its discredited competitors.

As I say, there are few Bayesians who can resist the inductivist allure of confirming evidence, of probabilistic support. It is quite standard for Bayesians to set down the usual definition (given in 3.3 above) of the support function $s(h, e)$ as $P(h|e) - P(h)$, the amount by which the probability of h is raised by the appearance of the evidence e, and to place some epistemological value on it. It mystifies me why they do. Bayesians are not supposed to be inductivists, and they should be as little interested as are falsificationists in how a hypothesis can be endowed with empirical support. I have been especially staggered by the energetic Bayesian response to the theorem of Popper and Miller (1983; 1987); the fervour of the comments of Eells (1988), Gaifman (1985), Good (1984; 1985; 1987; 1990), Howson (1989; 1990), Howson and Franklin (1985), Howson and Urbach (1989, 264–67), Jeffrey (1984), all Bayesian voices raised in protest at our interpretation of this result, certainly suggests that it really matters to some Bayesians that it should be possible for the unexamined content of a hypothesis to be supported by the evidence. For the life of me, I cannot see why it should matter. For pure Bayesians there should be no conclusion concerning the merit of a hypothesis to be drawn from the fact that it has been empirically confirmed (that is, its probability has gone up). This is not just because high levels of support are always in danger of being reversed under the pressure of unfavourable evidence; or that the support function has a number of frighteningly counterintuitive properties if understood as a measure of some valued aspect of a hypothesis relative to some evidence (see the references at the end of 3.3). It is more basic than that. A true Bayesian should simply not be interested in whether a hypothesis is supported or not. If he is interested, he has turned surreptitiously into an inductivist. About that no more need be said.

Its failure to nominate any aim for science as a whole means that although there is a Bayesian epistemology, a theory of how an agent's degrees of belief are related, there is really nothing that can be described as Bayesian methodology. Only one definite rule or procedure can be found residing properly within the Bayesian framework, and that is Bayes's rule, the exhortation to maximize expected utility. In Chapter 7 I shall look critically at a justly celebrated attempt to use this rule to explain why scientific investigators typically take account of all the evidence at their disposal, and (subject to cost) make an effort to gather more evidence. My conclusion will be that the explanation is in several respects unsatisfactory; and that, in any case, Bayes's rule can hardly be accepted as a rule of rational decision making. Even if that conclusion is invalid, it would be fanciful to claim that the whole of methodology can be reduced to applications of Bayes's rule. In practice, Bayesians make no such claim. What they do is to show that standard falsificationist methodological prescriptions can be redescribed, apparently more precisely, using probabilistic language; and this redescription is then, without the least indication that true deductivist principles are being abandoned, taken to amount to a Bayesian explanation (usually a more precise explanation) of the methodological rule in question. But a careful look shows that no explanation is actually supplied.

Take, for example, the familiar methodological rule that enjoins us to prefer severe tests to relaxed ones. We may cheerfully grant that the idea of a severe test can be translated into probabilistic terminology. A test is severe if it looks for a phenomenon that is highly probable given the hypothesis under test, but highly improbable given background knowledge alone. (See Popper 1963, Addendum 2.) What comes out at the other end of the calculations (spelt out by Grünbaum in his 1976b, 108f, for example) is that a severe test has a much greater power than does a relaxed one to raise dramatically the probability of the hypothesis under test; that is, it is capable of providing much greater support. But that in itself gives us no indication that we should take the trouble to perform severe tests. Methodologically the Bayesian elaboration of the doctrine of severe tests is utterly empty. The same could be said about Bayesian resolutions of Hempel's paradox. Perhaps Bayesianism can show that, the world being the way it is, a white shoe gives less support to 'All ravens are black' than a black raven does (Earman 1992, 69–73). What it does not explain are the sampling techniques used to test simple empirical laws. The same applies also to the interminable current controversy (discussed in Howson and Urbach 1989, 270–284, and by many others) over the question of whether a hypothesis can be supported by evidence that was known before the hypothesis was suggested. Here I agree with both sides. A hypothesis can be supported by evidence that it merely accommodates. On the other hand, such support in no way rebounds to the credit of the hypothesis under test. But in this respect it is, in my view, no different from empirical support obtained in any other way. Empirical evidence is not used to support hypotheses, but to refute them.

A final, in its way richly instructive, example is the ingenious attempt of Dorling (1979) to solve the Duhem/Quine problem in a Bayesian manner (see Earman *op. cit.*, 83–85). The issue is this: we have a hypothesis h that taken alone is not testable; in conjunction with an auxiliary hypothesis k, however, it is testable. We test it, and it is refuted by the evidence e. Since it may be k that is responsible for the refutation rather than h, it appears that there is nothing that we can conclude about *h*. So, it may be asked, what is the point of the test? Dorling replies that, depending on the comparative values of the probabilities of *h* and *k*, and the probability of *e*, it may well turn out that *e* has a startlingly asymmetric effect on *h* and *k*. He considers the following example: *h* is Newtonian theory; *k* is some suitable agglomeration of auxiliary hypotheses concerning the methods of measurement employed; and the evidence *e* is the observed value for the secular acceleration of the moon, which contradicted the value computed by Adams on the basis of *h* and *k*. The example assumes that *h* and *k* are probabilistically independent of each other, and sets $P(h) = 0.9$, $P(k) = 0.6$; the three likelihoods $P(e|k\text{-\&-not-}h)$, $P(e|h\text{-\&-not-}k)$, and $P(e|\text{not-}h\text{-\&-not-}k)$ are set equal to 0.001, 0.05, and 0.05 respectively. Then $P(h|e) = 0.8976$ whilst $P(k|e) = 0.003$. It appears that Newtonian theory is hardly affected by the reported value for the moon's acceleration (its probability goes down by only 0.0024), whilst the auxiliary hypothesis *k* is badly compromised (its probability goes down from 0.6 to 0.003). Striking, I agree, even though I cannot take the numerical values very seriously. But what is supposed to be the significance of these calculations? The test will not, on Bayesian grounds, lead to rejection of the hypothesis *k*; for Bayesians, being unconcerned with truth and falsity, do not talk—and sometimes affect not to understand talk—of acceptance and rejection. And clearly if the example proves anything it proves too much. For if it really is possible to compute the probability $P(h|e)$ of Newtonian theory *h* on the evidence *e*, it is hard to see what is or ever was the need for the auxiliary hypothesis *k*, except as an ancillary—a kind of arithmetical catalyst—to the calculation. In other words, the Bayesian analysis is at a loss to explain a perfectly uncontroversial feature of mature science: the employment of auxiliary hypotheses in order to bring a hypothesis face to face with the world.

The next chapter will consider the question of the use of pointed empirical investigation in science, and whether Bayesianism can explain it as an application of Bayes's rule. Then in Chapter 8 I shall have a few words to say about Bayesianism's claim to represent adequately scientific objectivity; that is, I shall look askance not only at its methodological ineffectiveness, but also at its failure to describe correctly what is going on. But the principal point has already been made, more than once. Bayesianism voids scientific activity of its true purpose, and reduces the study of human knowledge to the study of how beliefs change, rather than to the study of how they are to be changed. All in the name of deductivism too.

Technical Note

The argument in the fourth last paragraph of 6.4 may be set out more formally as follows.

P is Musgrave's epistemic inductive principle, 'It is reasonable to accept a theory that has survived severe critical scrutiny'. H is some theory, and e is a report (which we may suppose to be known to be true) that H has been exposed to, and has survived, severe critical scrutiny. A abbreviates 'It is reasonable to accept'. The principle of A-transmission (not in dispute here) says that if AH, and H implies h, then Ah. The derivation

$$P, e \text{ therefore } AH \tag{1}$$

is deductively valid. If P is true, then of course AH is also true. But P is not known to be true, though perhaps it is reasonable to accept it. By the principle of A-transmission, if AP then AAH. Hence AAH. But we cannot in general conclude by means of (1)—together with the assumption AP—that AH, which is presumably the conclusion that Musgrave wants.

Now suppose that e^* is a report (which we may suppose to be known to be true) that P has been exposed to, and has survived, severe critical scrutiny. The derivation

$$P, e^* \text{ therefore } AP \tag{2}$$

is deductively valid. If P is true, then of course AP is also true. But P is not known to be true, though perhaps it is reasonable to accept it. By the principle of A-transmission, if AP then AAP. Hence AAP. That is all that the derivation (2) yields. Although of course we can conclude also that if AP then AP, we do not need to appeal to anything beyond the most trivial logic in order to do it. The looming infinite regress is obvious. That it is a vicious regress is, I hope, also obvious. The elements of the sequence . . . AAP, AP, P get longer and longer the further back we go, and there is nothing that could conceivably count as a starting point.

Let K be a name for the scheme 'if-AAH-then-AH'. If K is true, then by (1) if AP then AH. But unless it is analytic, K is not known to be true. Perhaps it is reasonable to accept it. By the principle of A-transmission, if AK then A(if-AP-then-AH). Even if we admit the principle that if A(if-H-then-K) and AH then AK, all we shall ever squeeze out of this is that if AAP then AAH. Nothing is going to banish that double AA short of the assumption (unjustified and unreasonable in Musgrave's eyes, I should have thought) that K is true.

7

On the Maximization of Expected Futility

7.1. Introduction

In the previous chapter we have seen how little genuine methodology, active involvement in the world, there is in the Bayesian philosophy. The most general methodological question of all is, I suppose, the question of why empirical science is practised at all, why investigators seek new evidence. My intention in this chapter is to argue that the approved Bayesian answer to this question is seriously defective: either it fails altogether or it explains too much, putting a value on empirical investigations that are demonstrably valueless. In contrast, the falsificationist answer—that we look for new evidence in order to falsify our theories—seems to me to be amazingly obvious, not needing to be spelt out in detail. Of course, it remains open to standard justificationist misgivings, such as those of Lieberson (1982; 1983) and Maher (1990, 103–05), but I can't help that. That falsificationism offers no "argument to show that the chance of getting closer to the truth by gathering evidence outweighs the risk of getting further away from the truth" (Maher *op. cit.*, 104) hardly amounts to a criticism of the doctrine that the gathering of evidence is an excellent (though fallible) way of exposing error.

The Bayesian explanation I shall be criticizing is due in various forms to Ramsey (undated/1990), Savage (1954/1972, Chapter 7), Öpik (1957), Lindley (1965, 66), and Good (1967/1983). It has recently been generalized from the standard case of Bayesian conditionalization to the probability kinematics of Jeffrey (1965/1983) by Graves (1989; see also Skyrms 1987a, 1990). In my opinion the explanation is already sadly adrift in the standard case; but thanks to Graves's work it is now possible to make the point much more aggressively. Because my purpose is negative I shall be able to spare the reader nearly all the underlying algebra (which is neatly presented in several of the works just cited). All I shall do is give an example, which is elaborated in more than one way. In its final form, it is an example of two experiments Φ and Ψ, each of which reverses the effect of the other: after both experiments have been performed the agent is in exactly the epistemic state he was in before he started. But if Φ was worth doing the first time (as Bayesian principles say it was) it is worth performing again; and the same goes for Ψ, *ad ridiculum*. This is a new kind of scientific revolution in permanence.

7.2.   The Requirement of Total Evidence

The problem was vividly posed by Ayer (1957), and has been very well explained by Hilpinen (1970, Section 1). In the theories of probability advanced by Keynes (1921), Carnap (1950), and others, statements of probability of the form $P(h|e) = \lambda$ where λ is a numeral are "necessarily true, if they are true at all" (13). Suppose h is a hypothesis in which the agent has an interest; perhaps he intends to bet on it. Why, asked Ayer, should he prefer to evaluate, and place his bets according to, $P(h|e)$, where e is all the evidence in his possession, rather than $P(h|e-)$, where $e-$ is a fragment of e? All the associated probability identities are necessary truths, and it is not easy to see on Keynesian principles why any one of them should be superior to any other one. It is in response to just this problem, to be sure, that Carnap introduced the *requirement of total evidence*: "in the application of inductive logic to a given knowledge situation, the total evidence available must be taken as basis for determining the [probability]" (1950, 211). But "how can it possibly be justified on Carnap's principles?" (Ayer, 14). Ayer suggested that the only move available to Carnap "would be to make the principle of total evidence part of the definition of probability" (15). But this only draws attention to a further problem. For if e is the agent's total evidence then $P(h|e)$ is the value of his probability and that is that. What incentive does he have to change it, for example by obtaining more evidence than he has already? He might do so, enabling his total evidence to advance from e to $e+$; but in no clear way would $P(h|e+)$ be a better evaluation of probability than $P(h|e)$ was.

Although Ayer limited himself to theories of logical probability, his questions are equally pertinent to subjectivistic or personalistic Bayesianism. Why should an agent take all his evidence into account? If he does so, why should he exert himself to look for more evidence? Although statements of probability are no longer necessarily true, they are even more obviously on a par with each other. (One agent may have more evidence than another has, but, says personalistic Bayesianism, his probability distribution is not the worse for that. Hence there can be no reason to prefer an agent's probability distribution at one moment to the same agent's distribution at an earlier moment.) Indeed, Ayer's questions will arise for pretty well any measure $m(h|e)$ connecting hypothesis and evidence. An example of a measure for which they are easily answered is Popper's degree of corroboration $C(h, e)$. The insistence that "$C(h, e)$ can be interpreted as degree of corroboration only if e is *a report on the severest tests we have been able to design*" (Popper 1959a, 418) makes good sense, and it is clear what the advantage of further investigation—if properly directed—can be: the gain is not in the degree of corroboration, but in the stringency of the tests. Nonetheless the proviso is often ignored. Howson and Urbach, for instance, think that they are able to derive from Popper's proposals the consequence that "all . . . unrefuted theories . . . are equally 'corroborated'" (1989, 5).

7.3.   Good's Answer

Ayer asked for two things to be explained (or, in his jargon, justified): the requirement of total evidence, and the search for new evidence, for empirical activity itself. In view of the fact that both Skyrms (1987a) and Graves (1989) use the phrase 'principle of total evidence' when they intend the principle that new evidence should be sought, I wish to make it clear that I do not identify Ayer's questions. (They were first sharply distinguished by Hilpinen *op. cit.*) Moreover, the former— why should we not ignore some of the evidence in our possession?— must be distinguished also from a third question: why should evidence, if it is not ignored, be taken account of via Bayesian conditionalization? I stress that I have no comment to make about this third question, and shall simply grant the Bayesian requirement that, if a new statement e is accepted into the evidence between time s and time t, then the probability of h must change from $P_s(h)$ to $P_t(h) = P_s(h|e)$. In particular, $P_t(e)$ must surge to 1. But if the requirement of total evidence is not complied with, $P_t(e)$ may be less than 1.

There is a celebrated response to Ayer's two questions, originally due to Ramsey *op. cit.* and discovered independently and popularized by Good *op. cit.*, that revolves around Bayes's rule, the rule of maximizing expected utility. The rule may be formulated thus. Let $\Gamma_0, \Gamma_1, \ldots$ be the possible acts open to an agent, and $h_0, h_1, \ldots$ a family of exclusive and exhaustive hypotheses. Write $u[\Gamma_i, h_j]$ for the utility of performing Γ_i in the circumstances described by h_j, and $P(h_j)$ for the probability of h_j. Then the expected utility $E(\Gamma_i)$ of Γ_i is given by the formula

$$E(\Gamma_i) = \sum_j u[\Gamma_i, h_j]P(h_j) \tag{1}$$

Bayes's rule instructs the agent to perform that act Γ_i (or one of those acts if there is more than one) that maximizes the expected utility E. This rule is the core of orthodox Bayesian accounts of decision making. An example limited to two acts and two hypotheses is given in the next section.

Good showed that, if an agent is preparing to make a decision with the aid of Bayes's rule, then the expected utility of the double-decker act of first collecting further evidence (provided there is no cost) and then making the decision in the light of the newly accumulated evidence, will never be less than the expected utility of the simple act of charging ahead and deciding in the absence of further evidence. It is, therefore, always worth performing a free experiment and using the information obtained, as long as the experiment is what Skyrms calls salient (that is, it has some probability—in the agent's judgment—of affecting the eventual decision); and since evidence already in the agent's possession is free, it too is worth using; the requirement of total evidence, Good concludes, is a special case of the principle that free salient evidence should be collected and used.

Good's argument is incontestably lucid, but not incontestable; and in 7.5 I shall argue that it achieves less than is commonly supposed. (For other relevant comments, see Seidenfeld 1991, Section 3, and Seidenfeld and Wasserman 1991.) Let me now, however, get out of the way without delay two other possible explanations of the virtues of the requirement of total evidence.

The first, which to my knowledge has never been proposed, is the transcendental argument that if agents were permitted, whenever it suited them, to strike out an item of evidence, then betting would be reduced to a pointless activity. I dare say that this is so, though I question the aptness of the word 'reduced'. But it shows only that Bayesianism—if it is to make sense of betting—may need to make an explicit assumption of something like the requirement of total evidence. It does nothing to show that the requirement follows from more basic Bayesian teachings, and therefore that Bayesianism can explain it.

The second argument relies on the idea, embraced by Good (1968, 124), for instance, but emphatically repudiated by many Bayesians (for example, Howson and Urbach *op. cit.*, 54–56), that probabilities are subjective estimates of something objective, something like logical probabilities. If successful, this argument would go some way towards dealing also with Ayer's second question, since it would give some explanation of why it is worth the agent's while to collect new evidence. On such an understanding of probabilities, there might seem to be some merit in the claim that the better informed decision—the decision based on the total evidence—is the better decision. I disagree. It might perhaps be so if there were much truth in one of Bayesianism's favourite doctrines—that in the presence of mounting evidence, probability distributions converge. But this supposed convergence of opinion, though demonstrable in some elementary cases, is unobtainable in general. I discuss this in Chapter 8 below.

Before we proceed to a criticism of Good's argument, it may be helpful to look at a simple example.

7.4. Acquiring a Necktie

The agent urgently needs a necktie to wear with his only suit. Only one tie is available, but he does not know whether it will suit his suit (h) or clash with it (not-h). He does not even know what colour the tie is, though he believes with probability 0.6 that it is blue (e) and with probability 0.4 that it is green (not-e). The options open to him are to hire the tie (act Γ) or buy it (act Δ). The probabilities $P(x|z)$ and utilities $u[A, a]$ are given in these tables:

$P(x|z)$	z	e	not-e
x			
h		0.9	0.4
not-h		0.1	0.6

$u[A, a]$	a	h	not-h
A			
Γ		20	10
Δ		40	-30

The agent's preferred option is to buy a matching tie, and the next best thing would be to hire one. Hiring an incongruous tie would be worse, but worst of all would be being tied to a shocker in perennity. The reader is invited to find the probabilities plausible. We may easily deduce from them that $P(h) = P(h|e)P(e) + P(h|\text{not-}e)P(\text{not-}e) = 0.9 \times 0.6 + 0.4 \times 0.4 = 0.7$.

It is straightforward to calculate the expected utilities:

$$E(\Gamma) = u[\Gamma, h]P(h) + u[\Gamma, \text{not-}h]P(\text{not-}h)$$
$$= 14 + 3 = 17$$
$$E(\Delta) = u[\Delta, h]P(h) + u[\Delta, \text{not-}h]P(\text{not-}h)$$
$$= 28 - 9 = 19 \tag{2}$$

The maximum expected utility, which we shall write $E(\Gamma/\Delta)$, is thus 19, and buying the tie is the act recommended by Bayes's rule.

The agent has the opportunity to perform an experiment Ω that will reveal whether the tie is blue or green; no more. Let us suppose that he does perform it and that the result is e: the tie is blue. If he revises his probabilities by Bayes's rule, he obtains the following expectations:

$$E(\Gamma, e) = u[\Gamma, h]P(h|e) + u[\Gamma, \text{not-}h]P(\text{not-}h|e)$$
$$= 18 + 1 = 19$$
$$E(\Delta, e) = u[\Delta, h]P(h|e) + u[\Delta, \text{not-}h]P(\text{not-}h|e)$$
$$= 36 - 3 = 33 \tag{3}$$

The maximum expected utility, written $E(\Gamma/\Delta, e)$, goes up to 33, and the decision is still to buy. On the other hand, if it turns out that the tie is green (not-e) then the expectations become:

$$E(\Gamma, \text{not-}e) = u[\Gamma, h]P(h|\text{not-}e) + u[\Gamma, \text{not-}h]P(\text{not-}h|\text{not-}e)$$
$$= 8 + 6 = 14$$

$$E(\Delta, \text{not-}e) = u[\Delta, h]P(h|\text{not-}e) + u[\Delta, \text{not-}h]P(\text{not-}h|\text{not-}e)$$
$$= 16 - 18 = -2 \qquad (4)$$

There is a sharp fall in expectation: $E(\Gamma/\Delta, \text{not-}e)$ is only 14. Moreover, the preferred decision switches to hiring the tie. The experiment Ω is therefore what Skyrms calls salient.

The expected value $E(\Gamma/\Delta, \Omega)$ of the double-decker act of deciding between Γ and Δ after probabilities have been conditionalized on the outcome of Ω is given by

$$E(\Gamma/\Delta, \Omega) = E(\Gamma/\Delta, e]P(e) + E(\Gamma/\Delta, \text{not-}e]P(\text{not-}e)$$
$$= 33\times0.6 + 14\times0.4 = 25.4 \qquad (5)$$

Since $E(\Gamma/\Delta, \Omega)$ is greater than $E(\Gamma/\Delta)$, the act of performing Ω, then choosing between Γ and Δ is preferred by Bayes's rule to the act of choosing between Γ and Δ without benefit of Ω. That this is true for every salient experiment is the incontrovertible content of the theorem of Ramsey and Good.

7.5. Criticism of Good's Answer

There is nothing wrong with the algebra. Nonetheless, there are to my mind at least four things badly wrong with Good's argument. First of all, the argument as it stands simply does not impinge on the question of why evidence is collected in those innumerable cases in theoretical science in which no decisions will be, or anyway are envisaged to be, affected. Perhaps this fault could be repaired by resort to epistemic utilities (say, in the manner of Levi 1977). Perhaps not; epistemic utilities do not blend well with many versions of Bayesianism (see, for example, Rosenkrantz 1977, 147f, or Harsanyi 1985). In the second place, the argument is really much too general. If it is valid, then it recommends not only the gathering of evidence, but any other activity (such as the consumption of alcohol) that, in the agent's opinion, might alter probabilities sufficiently to affect the resulting decision. The third difficulty is that Good's proof of the principle that free salient evidence should be collected is invalid: it works only if the requirement of total evidence has already been adopted. Good himself noticed this gap in his proof and tried to plug it (others have managed to slide over it without remark). I shall explain the problem in the next paragraph. A fourth difficulty is a consequence of the third: since Good derives the requirement of total evidence as a corollary of the principle that free salient evidence should be collected, nothing is proved: like all proofs, the proof of the requirement of total evidence is circular, but unlike some proofs it constitutes a circle of very small radius.

The third difficulty, the central one, is this: the desire to perform acts of greatest possible expected utility does indeed, as Good showed, encourage us to collect new evidence; but when this new evidence is at last in our possession it may turn out that the maximum available expected utility is less than it was before we began. This is illustrated by the example above: if not-e obtains, the necktie being green, then the expectation decreases from $E(\Gamma/\Delta)$ to $E(\Gamma/\Delta, \text{not-}e)$, from 19 to 14. Conditionalizing on not-e blackens the agent's day. It would seem that in these circumstances the devout Bayesian, unless committed to heeding all the evidence, should be ready to ignore the evidence not-e. Indeed, the composite act A that consists of performing Ω, conditionalizing on the evidence if e obtains and ignoring it if not-e obtains, and then choosing between Γ and Δ, has the highest expected value of all acts considered:

$$E(A) = E(\Gamma/\Delta, e)P(e) + E(\Gamma/\Delta)P(\text{not-}e)$$
$$= 33 \times 0.6 + 19 \times 0.4 = 27.4. \tag{6}$$

Is the composite act A available to the agent? Good does not explicitly mention it, but he implicitly rules it out in what I can describe only as an ad hoc manner. This is done by assuming that the agent never gets embroiled in "a perpetual examination of the results of experiments, without a decision" (1967, 321; 1983, 180). It can be agreed that if the agent genuinely does disregard the new evidence he is back exactly where he started, and Bayes's rule once more enjoins him to seek it out; and so on to eternity. But this assumption of Good's really will not do. In the first place, it is not even the right assumption: what must be ruled out are not acts that never lead to decisions but decisions that never lead to acts. That Good was not clear about this may be seen from his initial formulation of the crucial assumption (1967, 319; 1983, 178):

> Suppose that we have r mutually exclusive and exhaustive hypotheses, h_1, $h_2, \ldots h_r$, and a choice of s acts, or classes of acts, $A_1, A_2, \ldots, A_s$. It will be assumed that none of these classes of acts consists of a perpetual examination of the results of experiments, without ever deciding which of $A_1, A_2, \ldots A_s$ to perform.

But there can surely be no need to exclude acts so weird as to consist in part in their own unperformance. What needs to be excluded, if the proof is to succeed, is the possibility that use of Bayes's rule could lead to an unending sequence of decisions concerning experimentation, effectively preventing the agent from proceeding to the second level of the double-decker decision process, namely that of deciding between the acts $A_1, A_2, \ldots, A_r$. And here another difficulty arises. What on earth in Bayesian principles allows one to suppose that this is not a possibility? Who is to say that it is not an unintended consequence of the use of Bayes's rule in these double-decker processes? In my view, that is precisely what it is: the requirement of total evidence, far from being a consequence of Bayes's rule, is not even consistent with it. The hard truth is that the unrestricted use of Bayes's rule does lead to vacillations of the kind that Good detected. In Jeffrey's probability kinematics, I shall argue in 7.6, this failing becomes almost impossible to conceal.

M. Bernard Walliser has suggested to me that an agent who decides not to recognize incoming evidence, and consequently does not update his probabilities in its light, is violating the principle of (causal or probabilistic) independence of the hypotheses h and not-h with respect to the acts Γ and Δ. (I am paraphrasing a personal communication.) But, it seems to me, quite the same holds when the agent behaves in an orthodox manner, swallows the data, and conditionalizes. The probability of h depends on which of the acts the agent performs; that is, whether he performs Γ/Δ or Ω and then Γ/Δ. In any case, there are many examples where the decision taken affects the probabilities; an instructive one is to be found in Jeffrey 1965/1983, 8f. I cannot think of any good Bayesian excuse for interdicting either these acts or acts such as A.

All the Bayesian can do in the face of this contradiction between the requirement of total evidence and Bayes's rule is to drop the attempt to explain the former by the latter, and to temper the latter to the demands of the former; that is, to insist that all expectations compared in any application of Bayes's rule be calculated in conformity with the requirement of total evidence. This is perhaps less restrictive than Ayer's suggestion to make a dependence on total evidence "part of the definition of probability"; for though it gives as little freedom to make a decision as does Ayer's proposal (the Bayesian agent never gets the chance to make a real decision, let alone a rational one), it allows the possibility, desired by several Bayesians (Skyrms 1987b; Howson and Urbach 1989, 270–75), of sensible talk about subjunctive distributions (that is, distributions were-the-evidence-other-than-it-is). That we are able to, and sometimes—for example, in courts of law—are obliged to disregard relevant evidence at our disposal is, I suppose, not in dispute.

If the requirement of total evidence is, in one way or the other, taken for granted—as it will be for the remainder of this chapter—then it may seem that Bayesianism can record a genuine success. (Skyrms [1990, 247f] lauds the argument as "a beautifully lucid answer to a fundamental epistemological question".) For Ayer's main objection to this modification—certainly the main objection that could worry a personalist—was that it only shifted the problem to that of explaining why we should seek further evidence; and this problem—now that the requirement of total evidence is axiomatic—is solved by Ramsey's and Good's proof. But unfortunately the adequacy of an explanation hangs on more than a valid derivation. Unobjectionable premises are needed too; and Bayes's rule, when remoulded to accord with the requirement of total evidence, encourages—shall we say—scepticism. The example in 7.4 demonstrates that the upshot of indulging in the act that maximizes expected utility may be a situation in which expected utility is sensibly diminished: if Ω yields the result not-e then the agent's expectation is significantly below what it was at the start. Now there is nothing much wrong with this if expected utility is not regarded as a value in itself—an action may be methodologically advisable even if it later turns out that it achieved nothing. (The obvious example is the subjecting of a theory to a severe test. If the theory is not falsified by the test then we learn nothing. Nonetheless, the test was worth doing. For it is in its method, not in its products, that the rationality of science resides.) But if expected utility is not itself valuable—in the way that actual utility is—then it is not apparent to me on what grounds Bayes's rule is thought to be a rule of rationality.

To be sure, there may be no grounds. Bayes's rule may just be a postulate, a conjecture. I am not opposed to this way of looking at the matter—indeed, I greatly prefer it—provided that—like any other conjecture—the rule is searchingly criticized, and eliminated if found wanting. In the rest of this paper I shall do my best to show that when offered as an explanation in probability kinematics for the principle that new evidence should be collected, Bayes's rule leads to absurdity; that it is wanting; that it must be eliminated.

7.6.   Probability Kinematics

Evidence has so far been construed, as usually it is at an elementary level, to be certain, or at least to have unit probability; what happens when evidence is used is that a statement e that previously had a probability properly between 0 and 1 is allocated a probability equal to 1. Now this requirement can well be relaxed; and the unstable character of Bayes's rule is most illuminatingly shown by a consideration of the more general case in which uncertain evidence is permitted. It may surely be conceded that sometimes at least the subjective acquisition of evidence leads to no certification of any statement in the language; all that happens is that probabilities get changed, but not to 0 or 1. Let me recall the example 'Observation by Candlelight' in Chapter 11 of Jeffrey 1965/1983, in which the observation of a piece of cloth in unhelpful light changes the agent's probabilities that the cloth is blue or green or violet, but does not convince him that any of these colours, or any other, is the true colour. (This example is of course the source of the example in 7.4, which in 7.7 will return to the domain of probability kinematics.) Jeffrey recommends that in these circumstances probabilities be updated by a procedure (Jeffrey conditionalization) that generalizes Bayesian conditionalization. It has recently been shown by Graves (1989; see also Skyrms 1987a and Skyrms 1990, Section II) that Good's proof can be reproduced in the context of this more general probability kinematics: that is, that the expected utility of first performing a free experiment, even if the outcome yields no certainty, and then making a decision, is never less than the expected utility of making the decision without bothering to do the experiment. I shall illustrate this result in the next section, and use it to show that, even if the requirement of total evidence is conceded, the Bayesian response to Ayer's questions cannot be right.

Allow me, before going further, to stress that it is no part of my business here to defend Jeffrey conditionalization as an appropriate way of reacting to uncertain evidence. I wish to keep an empty mind on the subject. For some doubts about the need for any such conditionalization procedure, the reader is referred to Levi 1967, and to the ensuing discussions in Harper and Kyburg 1968 and Levi 1969. See also Diaconis and Zabell 1982.

As Jeffrey himself points out (*op. cit.*, 161), the generalized procedure of conditionalization that he proposes opens up the possibility of there being two sources of evidence that effectively cancel each other out; they might, for example, be the inspection of cloth by candlelight and its inspection by starlight; or canvassing the opinion of an optimist on some issue, and doing the same with a pessimist. More abstractly, let P_0 be the agent's prior distribution, and suppose that $P_0(e) = \lambda$. We imagine that the carrying out of an observation or experiment Φ leads the agent to assign probability μ to the statement e. (It is intended throughout that λ, μ, ν, α, β are numerals for different real numbers properly between 0 and 1.) Jeffrey conditionalization reckons the new distribution P_1 by the formula

$$P_1(h) = \mu P_0(h|e) + (1 - \mu) P_0(h|\text{not-}e). \qquad (7)$$

Note that $P_1(e) = \mu$. The agent now undertakes a qualitatively different observation or experiment Ψ that induces a further amendment in the probability of e, which now becomes ν. The new distribution P_2 is obtained from P_1 just as P_1 was obtained from P_0. Using the fact that the probability of e never reaches the extreme values of 0 and 1, and also the definition (7) of the distribution P_1, we therefore obtain

$$\begin{aligned}
P_2(h) &= \nu P_1(h|e) + (1 - \nu)P_1(h|\text{not-}e) \\
&= \nu P_1(h\text{-\&-}e)/P_1(e) + \\
&\quad + (1 - \nu)P_1(h\text{-\&-not-}e)/P_1(\text{not-}e) \\
&= \nu[\mu P_0(h\text{-\&-}e|e) + \\
&\quad + (1 - \mu)P_0(h\text{-\&-}e|\text{not-}e)]/\mu + \\
&\quad + (1 - \nu)[\mu P_0(h\text{-\&-not-}e|e) + \\
&\quad + (1 - \mu)P_0(h\text{-\&-not-}e|\text{not-}e)]/(1 - \mu) \\
&= \nu P_0(h|e) + (1 - \nu)P_0(h|\text{not-}e). \qquad (8)
\end{aligned}$$

Observe that

$$P_0(h|e) = P_1(h|e) = P_2(h|e) = \ldots . \qquad (9)$$

$P_2(h)$ is therefore going to be identical with $P_0(h)$ if $P_0(e) = \nu$, that is if $\nu = \lambda$. Of course, it may not be. Nonetheless $P_3(h|e)$, defined in the obvious way from P_2 when the experiment Φ is repeated, will equal $P_1(h|e)$. The net result of the two experiments Φ and Ψ is precisely nil. Yet a true Bayesian will be impelled by the yearning to perform actions of maximal expected utility to carry out Φ, and then, his distribution amended to P_1, to carry out Ψ, and then Φ again, and then Ψ, and so on. This is a perfect absurdity. That is, Bayesianism either explains too little or explains too much. If the proofs given by Good and Graves are acceptable, then Bayesian agents attach value to experiments that from an objective viewpoint are patently valueless; if they are not acceptable, Bayesianism has not explained the value of empirical evidence.

It is no answer to the above criticism to suggest that the canny agent, foreseeing endless experimentation—or overwhelmed after a while by a sense of *déjà vu*—will adjust his probabilities so as to break himself out of the circle. That is not the story I have told. The agent I have described really is in danger of switching the probability of e from λ to μ and back to λ in perpetuity. No true philosopher will, I think, fail to recognize this situation. Nor is it any answer then to suggest that in that case the endlessly repeated experiments must indeed be of value to the agent, as may be seen by the fact that each performance is endorsed by Bayes's rule. That would amount to saying that Bayes's rule explains the rationality of our rational actions by showing that they conform to Bayes's rule. Yet I suppose that Bayesianism is intended to have some explanatory power. If it is, after all, held to be a priori valid, then I for one have nothing to say against it.

7.7. Acquiring a Necktie (continued)

This section will apply the above results to the example of 7.4, and may be omitted without real loss of continuity.

The agent finds that experiment Ω—determining the colour of the necktie—is no longer permitted, and that the best he can do is to inspect the tie either by candlelight (this is experiment Φ) or by starlight (this is Ψ). Initially $P(e) = \lambda$. If the tie really is blue then inspection by candlelight will induce the agent to assign to e a probability of μ; inspection by starlight will induce him to assign a probability of λ. If however the tie is green, he will set $P(e)$ equal to α if he inspects the tie by candlelight and equal to β if he inspects it by starlight. I shall write $\Phi\mu$ to indicate that as a result of experiment Φ the agent adopts μ as the probability of e (this need not be a change of probability); $\Phi\alpha$, $\Psi\lambda$, $\Psi\beta$ will be used analogously. These assignments may be regarded as psychophysical effects; and it may even be assumed that the agent knows that in each experiment it is only the actual colour of the tie that is relevant to how $P(e)$ gets fixed—though of course he does not know quite in what way it is relevant (if he did, he would be able to work out the colour of the tie). As it happens, the tie is blue, though no one knows that. Initially experiment Φ is one that—by Graves's result—the agent will be eager to perform (provided it costs nothing), whilst Ψ, which will change nothing, is worth nothing. Once Φ is performed, however, with the result $\Phi\mu$, the probability of e changes from λ to μ, and now it is Ψ that offers prospects of greater expected utility whilst Φ (which has, of course, just been done) is valueless, and is known to be valueless. The result of performing Ψ is then $\Psi\lambda$; that is, to return $P(e)$ to λ, and the whole futile process begins again. In Bayesian terms each successive experiment is worth doing, even though they are valueless when taken two at a time.

The actual progress of the calculations below follows closely the more general algebraic treatment given by Graves (though I have used simpler notation). We take the same probabilities $P(x|z)$ and utilities $u[A, a]$ as before. As we have seen in (9), these probabilities do not change when the agent snaps by Jeffrey conditionalization into a new distribution P_2. We set

$$\lambda = 0.6; \, \mu = 0.3; \, \alpha = 0.8; \, \beta = 0.1 \tag{10}$$

Thus $P_0(e) = 0.6$, as before, and $P_0(h) = 0.7$, as before. We need to know also what initial probability the agent assigns to the experimental results $\Phi\mu$, $\Phi\alpha$, $\Psi\lambda$, $\Psi\beta$. According to formula (G) on page 322 of Graves *op. cit.*, it is a canon of rationality that

$$\mu P_0(\Phi\mu) + \alpha P_0(\Phi\alpha) = \lambda,$$
$$\lambda P_0(\Psi\lambda) + \beta P_0(\Psi\beta) = \lambda. \tag{11}$$

If this is so (I cannot contest it here) then, since $P_0(\Phi\mu) + P_0(\Phi\alpha) = P_0(\Psi\lambda) + P_0(\Psi\beta) = 1$, we have

$$P_0(\Phi\mu) = (\lambda - \alpha)/(\mu - \alpha) = 0.4; \, P_0(\Psi\lambda) = 1;$$
$$P_0(\Phi\alpha) = (\mu - \lambda)/(\mu - \alpha) = 0.6; \, P_0(\Psi\beta) = 0. \tag{12}$$

(If the agent initially attaches probability λ to e then he is certain that experiment Ψ will leave his opinions unchanged.) We need to compute a good many expectations, and the following abbreviations will be used:

$E_0(\Gamma)$	the initial expectation of act Γ
$E_0(\Delta)$	the initial expectation of act Δ
$E_0(\Gamma/\Delta)$	the maximum of $E_0(\Gamma)$ and $E_0(\Delta)$
$E_0(\Gamma, \Phi\mu)$	the expectation of Γ if the experiment Φ yields result μ
$E_0(\Delta, \Phi\mu)$	the expectation of Δ if the experiment Φ yields result μ
$E_0(\Gamma/\Delta, \Phi\mu)$	the maximum of $E_0(\Gamma, \Phi\mu)$ and $E_0(\Delta, \Phi\mu)$
$E_0(\Gamma, \Phi\alpha)$	the expectation of Γ if the experiment Φ yields result α
$E_0(\Delta, \Phi\alpha)$	the expectation of Δ if the experiment Φ yields result α

$E_0(\Gamma/\Delta, \Phi\alpha)$ the maximum of $E_0(\Gamma, \Phi\alpha)$ and $E_0(\Delta, \Phi\alpha)$

$E_0(\Gamma/\Delta, \Phi)$ the expectation of the experiment Φ followed by the decision between Γ and Δ

The initial expectations $E_0(\Gamma)$ and $E_0(\Delta)$ are as they were in (2): $E_0(\Gamma) = 17$ and $E_0(\Delta) = 19$. The conditional expectations $E_0(\Gamma, \Phi\mu)$ and $E_0(\Delta, \Phi\mu)$ are computed in the same way except that $P_0(h)$ is replaced by $P_1(h)$, the updated probability of h; Jeffrey conditionalization (7) puts this equal to $\mu P(h|e) + (1 - \mu)P(h|\text{not-}e)$, which equals $0.9 \times 0.3 + 0.4 \times 0.7$; that is, 0.55. Hence

$$E_0(\Gamma, \Phi\mu) = u[\Gamma, h]P_1(h) + u[\Gamma, \text{not-}h]P_1(\text{not-}h)$$
$$= 20 \times 0.55 + 10 \times 0.45 = 15.5$$
$$E_0(\Delta, \Phi\mu) = u[\Delta, h]P_1(h) + u[\Delta, \text{not-}h]P_1(\text{not-}h)$$
$$= 40 \times 0.55 - 30 \times 0.45 = 8.5. \tag{13}$$

Thus $E_0(\Gamma/\Delta, \Phi\mu) = 15.5$. To compute the same expectations under the condition $\Phi\alpha$ (which will not obtain, though the agent is not to know that) we need only substitute α for μ in the above; the new probability of h becomes $\alpha P(h|e) + (1 - \alpha)P(h|\text{not-}e)$, which is 0.8. Thus

$$E_0(\Gamma, \Phi\alpha) = 20 \times 0.8 + 10 \times 0.2 = 18$$
$$E_0(\Delta, \Phi\alpha) = 40 \times 0.8 - 30 \times 0.2 = 26, \tag{14}$$

which means that $E_0(\Gamma/\Delta, \Phi\alpha) = 26$. Finally we must compute the expectation of the double-decker act. This is given by the weighted average of $E_0(\Gamma/\Delta, \Phi\mu)$ and $E_0(\Gamma/\Delta, \Phi\alpha)$ where the weights are respectively $P_0(\Phi\mu)$ and $P_0(\Phi\alpha)$. Thus

$$E_0(\Gamma/\Delta, \Phi) = E_0(\Gamma/\Delta, \Phi\mu)P_0(\Phi\mu) + E_0(\Gamma/\Delta, \Phi\alpha)P_0(\Phi\alpha)$$
$$= 15.5 \times 0.4 + 26 \times 0.6 = 21.8. \tag{15}$$

The agent's expectation thus increases from 19 to 21.8; he expects more from first doing Φ and then deciding between Γ and Δ (first looking at the tie by candlelight, then deciding whether to hire it or buy it) than from deciding at once. In Bayesian terms the game is worth the candle. On the other hand, there is no point in performing Ψ, in inspecting the tie under starlight. Similar but simpler calculations confirm that $E_0(\Gamma/\Delta, \Psi) = 19$; this is because it is already known that Ψ has no effect.

Let us now suppose that Φ has been performed and has yielded $\Phi\mu$: the agent has adopted μ as the new value of $P(e)$, and the distribution P_0 is replaced by P_1. The agent will now be moved to think about experiment Ψ (there is nothing to be gained from repeating Φ), and can do all the calculations again from the new starting point. The details are omitted. Suffice to say that (as Graves's result promised) Ψ is worth doing. $E_1(\Gamma/\Delta) = E_0(\Gamma/\Delta, \Phi\mu) = 15.5$, whereas

$$E_1(\Gamma/\Delta, \Psi) = E_1(\Gamma/\Delta, \Psi\lambda)P_1(\Psi\lambda) + E_1(\Gamma/\Delta, \Psi\beta)P_1(\Psi\beta)$$
$$= 19{\times}0.4 + 26{\times}0.6 = 16.3, \tag{16}$$

an increase, if only a modest one. Indeed, the expectation will be even further improved after Ψ has been performed and has yielded $\Psi\lambda$; it will rise from 16.3 to $E_1(\Gamma/\Delta, \Psi\lambda)$, which is nothing other than its original value of 19. At this point the agent will have gone full circle, and is ready to start again. Nietzsche would have been proud of him.

7.8. Conclusion

It may easily be checked that although each of Φ and Ψ in turn gives promise of enhanced expected utility, the combined experiment of Φ followed by Ψ offers nothing. It is not salient. In probability kinematics, that is to say, an experiment may not be salient, yet have a sub-experiment that is salient. Nothing as strange as this can happen in the special case delineated by Bayesian conditionalization: if there is some (subjective) probability that Ω will affect the eventual decision, then there is some probability that any extension of Ω will do the same. The difference is simply that, in the example given above, the outcome of experiment Φ actually determines the outcome of experiment Ψ—or, more accurately, both outcomes correspond in a unique manner to the colour of the tie; but nonetheless the agent, who does not know the specificity of this correspondence, is uncertain what will happen when Ψ is performed. No such ignorance is possible if an explicit report of the outcome of the experiment is incorporated into the total evidence.

Some may see this result simply as a criticism of the machinery of probability kinematics. To my mind, however, Φ and Ψ, as described above, are futile experiments. For neither of them can lead to any change in our assessment of the truth value of any proposition. If truth is our aim, as clearly I think that it is, not only in science but in all our investigations, there is nothing to be gained from experiments, even free ones, that have no bearing on truth. Some Bayesians, we have seen, think otherwise; and it can only be Bayes's rule, the recommendation to maximize expected utility, that makes them think so. This suggests that

Bayes's rule, far from being what Good *op. cit.*, 319, calls a "principle of rationality", is a very bad rule, and sometimes gives thunderingly bad advice. For (as I have stressed in g. of 2.2) it is not expected utility that the rational agent tries to maximize, but actual utility. How this is to be achieved is usually a matter of guesswork, to be sure; but that does not, as Bayesians and other justificationists think, make it an irrational thing to try. It is the method the agent adopts that primarily marks him out as rational or irrational, not the decisions that he makes. Recall that the field is rational decision-making, not rational-decision making.

8

Diverging Distributions

8.1. Introduction

In this chapter I conclude my assault on the Bayesian philosophy by criticizing one of its most hallowed doctrines—the doctrine that in the longish run differences of subjective opinion are typically suffocated by the overwhelming force of evidence, so that objectivity emerges as a kind of intersubjective agreement. But the question I shall consider, the question of whether probability distributions can evolve in a chaotic manner, has some independent interest, and my answer, that they can, may be challenging even to those who have no truck with subjectivist interpretations of probability. The question has two aspects—no doubt more—that I shall try to emphasize: *i*. Might increasing evidence force a probability distribution into random (or pseudorandom) fluctuations? *ii*. Might very small variations in a prior distribution lead to very large and uncontrollable variations in the posterior distribution? I have to say at once that, although I hope to offer some interesting and suggestive examples, I have no decently developed answers to these questions. But I shall try to say something about their significance. If I succeed in drawing attention to some little noticed possibilities, I shall be satisfied.

I shall start with a brief description of some of the elementary ideas of non-linear dynamics; then go on to some applications of these ideas in probability theory; and finally speculate on the consequences of these applications, and on what profounder results might be nearby.

8.2. The Logistic Function

It is now well established that dynamical systems that evolve deterministically may nevertheless evince the two non-classical features mentioned above: *i*. pseudorandom fluctuation and *ii*. extreme sensitivity to initial conditions. The first feature suggests that many of the processes we regard as random might, after all, be deterministically generated. The second assures us that, even if they are, they will not be predictable unless the relevant initial conditions can be ascertained with complete precision. In the terminology of Popper 1982a, Section 1, metaphysical determinism may, despite the appearance of randomness in the world, perhaps be true; but scientific determinism is not. I might remark that this seems to me to be one of those instances where a metaphysical doctrine that has wormed its way into science as a logical consequence of a genuinely falsifiable hypothesis must be rejected along with its host. (See also 4.3.d above.)

Both features, *i.* and *ii.*, are illustrated by the so-called logistic function. Let $f(t)$ be a real-valued function of a discrete variable defined by the following (non-linear) recurrence equation:

$$f(t+1) = \lambda f(t)(1 - f(t)) \tag{1}$$

where $f(0)$ is in $(0, 1)$ and λ is in $(0, 4)$. It is easily seen that the values of f stay in $(0, 1)$, since if y is in this interval then $y(1 - y)$ is in $(0, \frac{1}{4})$. For $\lambda \le$ 3 the function f behaves in a responsible manner, more or less rapidly settling down to a fixed point; in addition, the fixed point that f more or less rapidly settles down to varies smoothly and gently with the value of $f(0)$ and also with the value of λ. As Stewart observes, such a point attractor is "the least interesting from the point of view of dynamics . . . the long-term dynamics is to do absolutely nothing" (1989, 156f). As λ is increased beyond 3, things begin to change for almost all values of $f(0)$. At $\lambda = 3.2$ the eventual behaviour of f is to oscillate between two points; at $\lambda = 3.5$ or thereabouts, the oscillation has period 4; and so on until at 3.58 the function f ceases to simulate orderly behaviour and starts to jump around apparently at random. This is i. Note that there are a few exceptional cases; for instance if $\lambda = 4$ and $f(0) = \frac{3}{4}$ then f is constant. In addition, it is found, the way in which f changes varies very sensitively with the value of $f(0)$. Figure 1 shows the first 400-odd jittery steps f takes when $f(0) = 0.75000000012$ and $\lambda = 4$, in striking contrast to its imperturbability when $f(0) = 0.75000000000$. This is ii.

It is important to appreciate that the chaotic behaviour of f is not a mere whimsical oddity; it is as typical as homing in on a steady state or a stable limit cycle. And although the logistic function f does not, as far as I know, describe any real dynamical system, its behaviour is characteristic.

The recipe that Stewart (*op. cit.*, 146f) gives for chaos is: *stretch and fold*. To make a function f go chaotic, first enlarge it (say, by multiplying it by 4), then reduce the result in some different way (say, by multiplying it by the factor $1 - f$). This recipe is by no means foolproof, as the case of

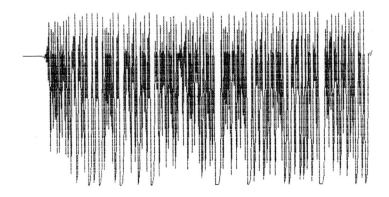

FIGURE 1

$\lambda \leq 3$ makes plain; but it is a good method to try. A process of stretching and folding that is even more primitive than the logistic function and again illustrates *ii.*—to a limited extent *i.* also—is this: Start with the decimal expansion of any real number, repeatedly multiply it by 10 (stretch), and take as the value of the operation the first digit after the decimal point (fold). Rational starting points will generate stable limit cycles, but irrational starting points will never do this. Yet every irrational number is as close as you like to a rational number and conversely. Rather as in Hadamard's 1898 result (cited in Part II, Chapter III, Section 3 of Duhem 1906 and in Section 14 of Popper *op. cit.*), closed orbits here are dense in eternal wanderings and vice versa.

This kind of discontinuity is exemplified in the case of the logistic function also. It is not the case that by moving $f(0)$ from 0.75000000012 closer and closer to 0.75 we would gradually obtain eventual behaviour that is closer and closer to the stationary state. (The value 0.75000000012 was chosen only because computers work to a limited number of significant figures.) I take it that a system is truly chaotic only when we have this kind of breakdown of what Popper calls the principle of accountability (*op. cit.*, Section 3). Let us refer to it briefly as hypersensitivity (to initial conditions).

Probability theory is never far away when chaos and randomness are being discussed; whatever else they do, we expect chaotic sequences to pass statistical tests for randomness. But I wonder whether there may not be other connections between probability and chaos; whether the probability function might not itself engage in chaotic behaviour. It seems worth considering. After all, the process of Bayesian conditionalization, which takes $P(h)$ to $P(h|e) = P(h)P(e|h)/P(e)$, looks very like a process of simultaneous stretching—dividing by $P(e)$—and non-proportional folding—multiplying by $P(e|h)$. I now proceed to investigate the possibilities.

8.3. The Binomial Distribution

The matter is not easy, so I should like to begin with a rather straightforward example, not itself involving conditionalization, that demonstrates how unbearably sensitive perfectly classical and familiar distributions may be to slight variations in initial conditions; or, more exactly, to slight variations in the parameters that characterize them.

The binomial distribution is characterized by a single parameter p, the probability of success, and gives values for the probabilities of r successes in n mutually independent trials. Coin-tossing is a simple example; a successful trial is an occurrence of Heads. Let p be the probability of Heads and X be the outcome of obtaining 500,000 ± 1,000 heads in a million tosses of a coin. If $p = 0.5$, then $P(X) = 0.9545$ (Feller 1968, 187). Likewise, if $p = 0.502$, the probability that the number of Heads is in the ínterval 502,000 ± 1,000 (which is effectively disjoint from the first one) is 0.9545; and in this case, by the symmetry of the distribution, it may be seen that $P(X) < 0.0228$. Hence a change of 0.002 in the value of $p = P(\text{Heads})$ engenders a change of at least 0.9317 in $P(X)$. It is obvious that such inconsistencies can be made as dramatic as desired by increasing the size of the sample taken.

It must be emphasized, however, that there are no violations of continuity in this example. If p is in the interval (0.5, 0.502) then $P(X)$ will be between the two values indicated above; and the closer p is to 0.5, the closer $P(X)$ will be to 0.9545. Thus there is no breach of the principle of accountability. To put the matter crudely, the quantifiers are the wrong way round. What is true is that *for (almost) every event Y* there is a number ϵ such *that $P(Y/p = 0.5 + \epsilon)$ differs substantially from $P(Y/p = 0.5)$*. What we seem to need for genuine chaos is—at least—something like this: *for every number ϵ there is an event Y such that $P(Y/p = 0.5 + \epsilon)$ differs substantially from $P(Y/p = 0.5)$*. Not quite that perhaps, but something on those lines.

I shall return to this example in 8.7. For the moment, I need say only that it illustrates the point, perhaps a pretty obvious one, that a small difference between the parameters of two probability distributions does not imply that the distributions themselves are in good agreement with regard to specific events. This phenomenon—from which it follows that the consequences of approximately true hypotheses may not be approximately true—should be familiar to all those who have followed the debate on verisimilitude. An excellent discussion of this point is to be found in Section 2 of Weston 1992.

8.4. A Chaotic Distribution

We turn to our main question—the evolution and possibly chaotic long-term behaviour of the function $P(h)$ for some event or proposition h. For present purposes it is not important how, or whether, the function P is interpreted; it may be a relative frequency, a propensity, a logical or subjective probability, or something else. What is important is that there should be an unlimited supply of events (or propositions) e_0, $e_1, e_2, \ldots$ such that the set $\{h\} \cup \{e_i \mid i \in \omega\}$ is completely independent. Thus h could be the set of throws with a coin under fixed conditions, and the e_i various restrictions on classes of coin throws (attributes, in von Mises's jargon); or h could be the proposition that God exists, the e_i reports of distinct worldly manifestations of good and evil. In each case I shall refer to the e_i as items of evidence, and shall, for simplicity, assume that the only issue for each i is whether e_i or not-e_i obtains. The simplest form of Bayes's theorem, linking the prior probability $P(h)$ with the posterior probability $P(h|e)$ of h on evidence e, is not contentious:

$$P(h|e) = P(e|h)P(h)/P(e). \tag{2}$$

If the evidence e is accumulated one item at a time, we may think of the values $P(h)$, $P(h|e_0)$, $P(h|e_0\text{-\&-}e_1)$, $\ldots$ as plotting the evolution of a function H in accordance with the recurrence relation

$$H(t+1) = P_{t+1}(h) = \lambda(t)P_t(h) = \lambda(t)H(t), \tag{3}$$

where $\lambda(t)$ is the term $P_t(e|h)/P_t(e)$. Under what conditions, if any, can we expect H either *i.* to fluctuate in a pseudorandom manner or *ii.* to depend hypersensitively on $H(0)$?

The popular wisdom, if it deserves to be so described, is that both *i.* and *ii.* are impossible. The view seems to be that, as evidence increases in quantity, even initially quite disparate distributions will converge to a common limiting distribution. I shall attempt to show that neither *i.* nor *ii.* is ruled out. We may note in passing also that stable limit cycles can exist: it is possible for distributions to go into a spin, never settling on any one thing. I therefore contest the popular view that, even if we start with different opinions, we shall eventually more or less agree. On the contrary, we may start with almost the same opinion, yet after a while disagree not only among ourselves but with ourselves.

Let us start with a consideration of just one item of evidence e. We wish to construct a distribution in which the value of $P(h|e)$, given above in (2), depends hypersensitively on $P(h)$. In general, given $P(h)$, the values of the likelihoods $P(e|h)$ and $P(e|\text{not-}h)$ are independently specifiable, and $P(e)$ is determined by the equation

$$P(e) = P(e|h)P(h) + P(e|\text{not-}h)P(\text{not-}h). \tag{4}$$

In order to avoid rather trivial constructions (similar to the second example of stretching and folding in 8.2), we shall require that these likelihoods be continuous functions of $P(h)$. So although some interesting effects can be produced by, for example, making $P(e|h)$ depend on the fifth digit in the decimal expansion of $P(h)$, we shall not pursue such eccentric distributions here.

It might seem that there exists an effortless recipe for producing a distribution that evolves in a truly chaotic manner. All that we need do is mimic the logistic function in its simplest form—that is, fix the likelihoods $P(e|h)$ and $P(e|\text{not-}h)$ at each stage in the evolution in such a way that

$$P(h|e) = 4P(h)(1 - P(h)). \tag{5}$$

But it is not quite as easy as that. For, provided $P(h)$ is not zero, (5) is equivalent, by Bayes's theorem (2), to the equation

$$P(e|h) = 4P(e)(1 - P(h)), \tag{5a}$$

while, by (4),

$$P(e|\text{not-}h) = P(e)[1 - 4P(h)(1 - P(h))]/(1 - P(h)). \tag{5b}$$

In addition to (5a) we therefore need to say something about $P(e)$. What we cannot do is what would be the simplest thing to do; that is, to give $P(e)$ the constant value ½. We would then obtain

$$P(e|h) = 2(1 - P(h)), \tag{5c}$$

$$P(e|\text{not-}h) = [1 - 4P(h)(1 - P(h))]/2(1 - P(h)). \tag{5d}$$

But, given that probabilities cannot exceed 1, from each of (5c) and (5d) follows some restriction on the value of $P(h)$. For example, it follows from (5c) that $P(h) \geq \frac{1}{2}$. Yet if $P(h)$ at any stage is large enough (in fact, larger than $(2 + \sqrt{2})/4$, which is approximately 0.853553391) then at the next stage it is transformed by (5) to a value that is less than $\frac{1}{2}$. In short, if $P(e)$ has to be fixed at $\frac{1}{2}$, (5) does not suffice on its own to define a suitable distribution, or any distribution at all.

There are a number of ways of getting round this problem. I shall consider one of them briefly in this section, another at greater length in the next. Neither sets $P(e)$ equal to a constant. One solution is to accept (5c) when $P(h) \geq \frac{1}{2}$, and to give the likelihood $P(e|h)$ a different specification when $P(h) < \frac{1}{2}$. The following allocation of likelihoods, in which for simplicity we take $P(e|h) + P(e|\text{not-}h) = 1$, does this.

$$\text{if } P(h) \geq \frac{1}{2} \text{ then } P(e|h) = 2(1 - P(h))$$

$$\text{and } P(e|\text{not-}h) = 2P(h) - 1$$

$$\text{if } P(h) \leq \frac{1}{2} \text{ then } P(e|h) = 2P(h).$$

$$\text{and } P(e|\text{not-}h) = 1 - 2P(h). \tag{6}$$

Note that (6) is already some way away from the simple idea encapsulated in (5). The likelihoods $P(\text{not-}e|h)$ and $P(\text{not-}e|\text{not-}h)$ are easily calculated using the addition law. Using the more explicit form of Bayes's theorem

$$P(h|e) = P(e|h)P(h)/[P(e|h)P(h) + P(e|\text{not-}h)P(\text{not-}h)] \tag{7}$$

that is obtained by expanding the denominator of (2) by the law of total probability, and writing y for $P(h)$, x for $P(h|e)$, and z for $P(h|\text{not-}e)$, it may be checked that generally for y in $(0, 1)$

$$\text{if } y \geq \frac{1}{2} \text{ then } x = 2 \quad y/(4y - 1)$$

$$\text{and } z = (2y - 1)y/(2 - 5y + 4y^2)$$

$$\text{if } y \leq \frac{1}{2} \text{ then } x = 2y^2/(1 - 3y + 4y^2)$$

$$\text{and } z = (1 - 2y)/(3 - 4y). \tag{8}$$

Simple calculations show that $P(h|e) = 0$ if $P(h) = 0$, and that $P(h|\text{not-}e)$ is indeterminate, approaching $\frac{1}{3}$ as $P(h)$ approaches 0. Likewise, $P(h|\text{not-}e) = 1$ if $P(h) = 1$, and $P(h|e)$ is indeterminate, approaching $\frac{2}{3}$ as $P(h)$ approaches 1. When $P(h) = \frac{1}{2}$, it may also be checked, $P(e|h) = 1$ and $P(e|\text{not-}h) = 0$. As in the logistic function, $\frac{3}{4}$ is a fixed point (and here $\frac{1}{4}$ is too): if $P(h) = \frac{1}{4}$ then $P(h|e) = P(h|\text{not-}e) = \frac{1}{4}$, and if $P(h) = \frac{3}{4}$ then $P(h|e) = P(h|\text{not-}e) = \frac{3}{4}$. That is, if $H(0) = \frac{1}{4}$ or $\frac{3}{4}$, then $H(t)$ will remain at that value in perpetuity. Figure 2 gives an example that compares such constant behaviour in 400 trials with highly deviant behaviour from an almost identical starting point, namely 0.250000001.

It has to be stressed that although this distribution is a decidedly unusual one, it is perfectly possible. If we think of $p = P(h)$ as a parameter of the distribution, indeed, we should be able to accept that all other probabilities are functions of $P(h)$. More than that: I want to claim that the formulas given under (6) determine a family of distributions differing only in a single parameter $P(h)$; and that the difference between the values assigned to $P(h)$ by two distributions marks the difference between those distributions. As observed at the end of 8.3 above, we cannot expect closely similar distributions in a family to assign closely similar probabilities to all individual events. It is the intensions, not the extensions, of the distributions of a family that must be similar to each other.

If you will swallow this for a single item of evidence e, there is no bar to your swallowing it for an indefinitely extended sequence of items e_0, e_1, ... But before I give an illustration I must mention one important complication, absent from the studies of deterministic chaos cited in 8.2, that affects the present discussion. The evolution of a probability

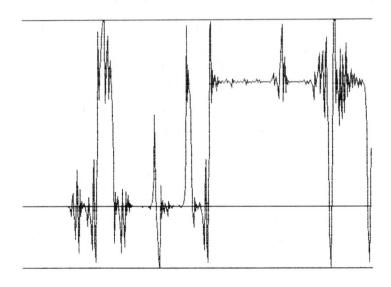

FIGURE 2

distribution is not self-contained in the way the evolution of the logistic function is: it depends on input, evidence. And evidence statements potentially have two values, true and false. It would not be unduly surprising (which does not mean that it would not be interesting) if under the influence of a particular sequence of evidence reports a probability distribution were to go chaotic. But to get anything like a thought-provoking result, we need to be able to suggest—I pretend no more—that chaos can be expected for typical sequences of evidence reports, $\pm e_0$, $\pm e_1$, . . . , where the $\pm$ is to be replaced either by nothing at all or by 'not-'. In what follows, unless otherwise stated, the sequence of $+$ and $-$ signs will have been randomly generated. But this does not imply that $P(e)$, which is given by (4), is constant and equal to 0.5. I shall return to this matter at the end of the next section.

8.5. Another Chaotic Distribution

A second way to generate a distribution that evolves chaotically is to forsake (5) altogether and instead to give the likelihood function $P(e|h)$ the opportunity to mimic the logistic function. As in (6) we set $P(e|\text{not-}h)$ equal to $1 - P(e|h)$.

$$P(e|h) = 4P(h)(1 - P(h)) \text{ and}$$
$$P(e|\text{not-}h) = 1 - 4P(h)(1 - P(h)). \tag{9}$$

The distribution given by (6) is, in effect, a linearization of that given in (9). In Figure 3 the parabola is $P(e|h)$ according to (9), the wigwam is $P(e|h)$ according to (6).

Again the likelihoods $P(\text{not-}e|h)$ and $P(\text{not-}e|\text{not-}h)$ are easily calculated. We have also by (4)

$$P(e) = 1 - 5P(h) + 12P(h)^2 - 8P(h)^3. \tag{10}$$

If $P(h)$ is in the open interval $(0, 1)$, the values of $P(h|e)$ and $P(h|\text{not-}e)$ are given in (11). As in (8) we write y for $P(h)$, x for $P(h|e)$, and z for $P(h|\text{not-}e)$.

$$x = 4y^2/(1 - 4y + 8y^2) \text{ and}$$
$$z = (1 - 4y + 4y^2)/(5 - 12y + 8y^2). \tag{11}$$

Simple calculations show that $P(h|e) = 0$ if $P(h) = 0$, and that $P(h|\text{not-}e)$ is indeterminate, approaching ⅕ as $P(h)$ approaches 0. Likewise,

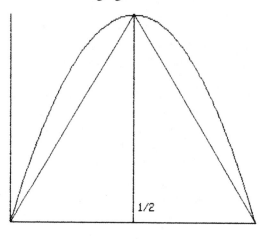

FIGURE 3

$P(h|\text{not-}e) = 1$ if $P(h) = 1$, and $P(h|e)$ is indeterminate, approaching ⅘ as $P(h)$ approaches 1. When $P(h) = ½$, it may also be checked, $P(e|h) = 1$ and $P(e|\text{not-}h) = 0$. The fixed points of $P(h)$ are not ¼ and ¾ as before, but $(2 \pm \sqrt{2})/4$: that is, 0.146446609 and 0.853553391 (approximately).

The values for $P(h|e)$ and $P(h|\text{not-}e)$ given in (11) do not look very symmetric. The symmetry is there, however, as the curves in Figure 4 reveal. The horizontal axis carries the value of y, ranging from 0 to 1. The curve that is convex at $y = ½$ (the deckchair-shaped curve) shows how z varies with y; the curve that is concave at $y = ½$ (the upper curve) shows how x varies with y. A similar, more angular, figure may be drawn for the distribution given by (6).

The general effect of conditionalization by e (rather than by not-e) is that values of $P(h)$ between the fixed points are increased, whilst those outside this interval are decreased. Conditionalization by not-e rather than e has the opposite effect. It may be shown that a long run of items of evidence all of the same sign will quickly force $P(h)$ arbitrarily close to one or other of the fixed points.

Let us set out again the position. We have a hypothesis h, and are about to collect a sequence of items of evidence, $\pm e_0, \pm e_1, \ldots$, each of which may be either positive (e_i) or negative (not-e_i). Whether they are positive or negative is a matter of chance here, though that is unlikely to be so in any real situation. (Only if the evidence is sufficiently varied will we produce the kinds of effects now to be documented.) We suppose

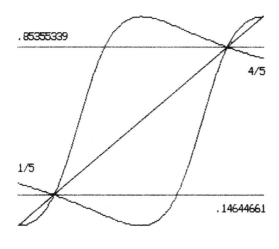

.85355339

4/5

1/5

.14644661

FIGURE 4

that the likelihoods for the positive and negative items of evidence are given. On receipt of each item, positive or negative, the value of $P(h)$ is updated by Bayes's theorem (2), yielding either $P(h|e)$ or $P(h|\text{not-}e)$. We may then take this posterior distribution as our new prior, drop its second argument, and start again. We follow the convention of (3) above, writing $H(t)$ for the value P has after the accumulation of t items of evidence, and plot it as a function of t.

Figures 5–8 depict four possible, and fairly typical, evolutions of the function $H(t)$. The horizontal line in each figure shows the (constant) evolution from $H(0) = 0.1464466094$. The serrated line shows an irregular, pseudorandom evolution from $H(0) = 0.14644662$. Each pseudorandom evolution is produced by a random sequence of 400 positive and negative items of evidence. These are represented respectively by 1s and 0s in the lists given in the technical note at the end of this chapter.

In Figures 6 and 7 $H(t)$ appears eventually to settle down very close to the fixed points $(2 - \sqrt{2})/4$ and $(2 + \sqrt{2})/4$ respectively. It is clear that these values are fairly powerful attractors, and that nearby values are positions of unstable equilibrium. But there is of course not the slightest reason to suppose that, if the evolutions were prolonged indefinitely, $H(t)$ would stay near these positions. This should perhaps be obvious from Figure 8, in which $H(t)$ remains within 0.0001 of the base line 0.1464466094 for a great deal of the first half of the performance (in

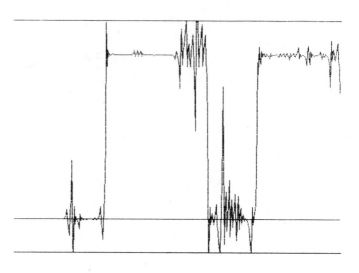

FIGURE 5

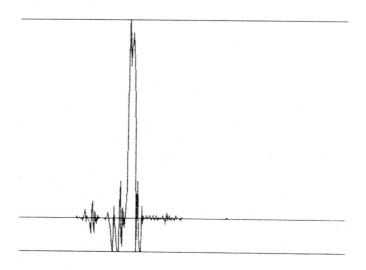

FIGURE 6

fact, for more than 150 of the first 200 trials), briefly rolls in a fine frenzy, calms down again, and finally increases dramatically.

The existence of round-off errors needs to be emphasized. In many examples of this kind, the calculated values of $H(t)$ genuinely stick at a

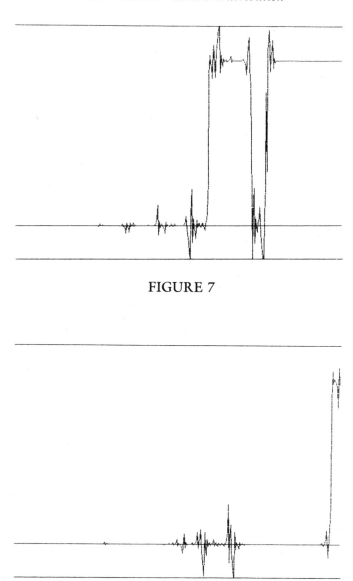

FIGURE 7

FIGURE 8

constant value in the neighbourhood of one or other of the (irrational) fixed points. Yet it is clear from (11), or from Figure 4, that $H(t+1)$ cannot precisely equal one of the fixed points unless $H(t)$ already equals it. Its appearing to do otherwise is an artefact of the computing apparatus, and means nothing. (On the machine used in the present investigation [a BBC Master with 65C102 co-processor], 0.1464466094 behaves like a fixed point, whereas 0.14644661 is not guaranteed to remain constant.) I have discarded all evolutions that degenerate in this manner. This selectivity doubtless explains the fact, which some alert readers may have noticed, that in each of the listings of outcomes in the Technical Note the number of 1s slightly exceeds the number of 0s. I would expect the preponderance to be reversed if $H(0)$ were set not near its lower fixed point but its upper one. Of the non-degenerate distributions I have seen, the examples here presented are, I think, completely characteristic.

It hardly needs to be said that the evolutions in Figures 5–8 give signs of both *i.* pseudorandom fluctuation and *ii.* extreme sensitivity to initial conditions. But whether they are truly chaotic is quite another question. "There is always the chance that the system is undergoing a very long and complicated transient state, before it settles down to some nice and regular behaviour" (da Costa and Doria 1992, 73). Tests for chaotic behaviour do exist (see, for example, Ruelle 1989, Chapter 9), but I do not know how they could be applied to examples such as these, or how much information they would yield. As da Costa and Doria (1991) have demonstrated, the problem of determining whether a system is chaotic or not is unsolvable in general. That does not mean that a firm decision may not be forthcoming in these cases involving Bayesian conditionalization. But it does indicate that obtaining the decision may be a thankless and extremely complex task.

Even if the technical possibility of a family of distributions such as that given by (9) is granted, we may well ask whether it is the kind of distribution that we might find in nature (if P is interpreted as a physical probability) or that someone might grade beliefs by (if P is interpreted subjectivistically). To this question I have no answer. There is at first sight something rather odd about a situation in which $P(e|h)$ depends explicitly on $P(h)$. Yet it is not impossible. Suppose, for example, that a hostess has invited to a splendid dance about twice as many guests as she can comfortably entertain. Then, amongst those who accept the invitation (h), the proportion $P(e|h)$ who think they made the right decision (e) could well depend, in a manner rather like that ordained by (9), on the proportion $P(h)$ of those who accept; if few, or most, of those invited accept, few of them will think that that was the right decision; if about half accept, most will be happy. Likewise, amongst those who do not accept, the proportion $P(e|\text{not-}h)$ who think they made the right decision will be at its lowest if $P(h)$ is about one half, and at its greatest if the dance is either overcrowded or deserted. To be sure, this is only a very roughly described distribution, but it may indicate that (9) is not utterly gilbertian.

It is certainly possible to design an urn model for the distribution described by (9). Let there be two connected urns, containing black and white buttons only. If urn A contains a proportion y of black buttons, then in some undisclosed manner urn C contains a proportion $4y(1-y)$ of black buttons. (This ought not to be excessively difficult to arrange; recall that if buttons are drawn at random in pairs from urn A then the proportion of draws that contain one button of each colour is $2y(1-y)$.) One button is drawn from each of the two urns. If the two drawn buttons are of the same colour then, in line with equations (11), urn A is reconstituted with a proportion $4y^2/(1-4y+8y^2)$ of black buttons; if the two drawn buttons are of different colours, on the other hand, then the proportion of black buttons in A is altered to $(1-4y+4y^2)/(5-12y+8y^2)$. In each case the composition of urn C is amended in accordance with the logistic function. For greater realism the reconstitution of the urns need not be thought of as an instantaneous process, nor is it necessary that the proportions of black buttons should be exactly the values stated. (If the values are exact, then the urns may have to be unimaginably capacious.) In this model the drawing of a button from C is the counterpart of obtaining an item of evidence. It is clear that the composition of urn A will exhibit the kind of chaotic evolution that has been illustrated above, and C even more so. I see no reason in principle why a physical model should not be built answering approximately to these requirements. But I am not here trying to boost (9)'s claim to serious attention as a realistic distribution. Although I should naturally be delighted if someone could discover it, or a relative, at work somewhere, its purpose here is only to indicate what might be possible.

One last point needs to be made about the value of $P(e)$. It is clear that neither (6) nor (9) is a distribution in which $P(e)$ takes the constant value ½. Yet we have generated the signs of the different items of evidence randomly; that is, with (objective) probability ½. (The point of this, it will be remembered, was to try to obtain typical data streams rather than untypical ones.) It might seem, therefore, that since the relative frequency of positive items of evidence approximates ½, $P(e)$ should do so too; and since the distributions that we have considered show no signs of learning, they can be disqualified on these grounds alone. But that is not so. Nothing in what has been said requires the different e_i to be in any respect similar; any more similar than are reports of the various tests of Newton's theory that involve behaviour of springs, pendulums, tides, planets, satellites, and Fletcher's trolleys. It is only for simplicity of exposition, indeed, that we have taken the likelihoods $P(e|h)$ and $P(e|\text{not-}h)$ to be the same at each stage of the evolution of the function H. Thus there is no reason to worry that $P(e)$ is not normally equal to ½. It would be good to have an example in which this additional condition was satisfied. But, I submit, it is not necessary.

8.6. Scientific Objectivity

One of the more extravagant pieces of apologetics of subjectivistic Bayesianism is that with growing evidence widely different probability distributions are almost bound to converge, thus securing a measure of agreement and even of objectivity. Dorling, for example, asserts boldly (1981, 120): "I do believe, as a good Bayesian, that if we all stayed around long enough and accumulated sufficient evidence then our posterior subjective probabilities would become arbitrarily close to each others'." The claim is made also on a number of occasions in Chapter 10 of Howson and Urbach 1989. On page 236, for example, they write: "The subjectivity of the premises of a Bayesian inference might suggest that the conclusion must be similarly idiosyncratic, subjective, and variable. Were this the case, the Bayesian approach would fly in the face of one of the most striking facts about science, namely its substantially objective character. However, it is not the case." They go on to quote Edwards, Lindman, and Savage (1963), who say: "More generally, two people with widely divergent prior opinions but reasonably open minds will be forced into arbitrarily close agreement about future observations by a sufficient amount of data." On pages 236–241 and 243–45 of Howson and Urbach's book (see also Earman 1992, 142f) some examples are presented in which limited convergence is an undeniable fact. At the close of the chapter, however, the perspective is decidedly more restrained: "perfectly sane scientists [we are now told] with access to the same information . . . normally approach a common view as the evidence accumulates"; and it is recognized that there is "the possibility of people whose predispositions either for or against certain theories are so pronounced and different from the norm that their opinions remain eccentric, even with a large volume of relevant data" (254). On page 290 fairly rapid convergence is said to happen "characteristically".

There undoubtedly do exist circumstances in which it is possible to demonstrate the convergence of probability measures modified by Bayesian conditionalization on the same pieces of evidence. Statisticians have generally been careful to state the conditions under which such theorems hold. Bayesian philosophers of science have sometimes been less meticulous. Despite endless trumpeting by them of the reality of convergence, I have never seen anything like a general proof of it. Nor have I ever expected to. (See Section 3(vii), on page 135 of my 1981.) And certainly uniform convergence, which is what would be needed to underwrite a claim of objectivity, seems an impossible dream. For it is perfectly obvious that, if the only restriction imposed on degrees of belief is that they should satisfy the probability axioms, it must be possible, however copious the evidence e is, for the agent's degrees of belief to be measured by any probability function assigning probability 1 to e. It must be possible, that is, for any hypothesis h independent of e to receive any probability in the unit interval; so unless it is deemed possible (on Heaven knows what grounds) for empirical evidence eventually to settle all questions conclusively, there is always the possibility of wide divergence. To be sure, if the sequence $\pm e_0, \pm e_1, \ldots,$ of evidence statements in some sense converges to h then it may be anticipated that $P_t(h)$ will converge to 1. Something like this seems to be the content of Doob's martingale convergence theorem (see Earman *op. cit.*, 144f). No uniform convergence (that is, the same rate of convergence for different initial values of $P(h)$) can be expected even here. In any case, it is only hypotheses at the lowest level (empirical generalizations) that can in this way be approximated by greater and greater wagonloads of evidence. A similar point is made by Good against what he calls the Second Induction Theorem (1975, 63; 1983, 165).

It should perhaps be added that de Finetti was clearer than most about convergence; even if it is often a fact (which I do not care to dispute), it is of no objective significance. On page 152 of his 1937/ 1964 we read that "there are rather profound psychological reasons which make the exact or approximate agreement that is observed between the opinions of different individuals very natural, but . . . there are no reasons, rational, positive, or metaphysical, that can give this fact any meaning beyond that of a simple agreement of subjective opinions".

Despite what I have said, de Finetti may be right about the naturalness of agreement. He is right too about its lack of significance: consensus is not correctness. It may be worth adding to this that correctness is not justified correctness.

The considerations of 8.4 and 8.5 above, if they have any validity, only reinforce these conclusions. For they suggest that agents who start off in almost complete (probabilistic) agreement may in special, perhaps unusual, circumstances come to diverge in their opinions more and more, despite being exposed to the same evidence. If this is so, it is scarcely astonishing that agents who start off at loggerheads might fail to end up in agreement.

8.7. Approximate Distributions

A rather obvious objection to the ideas of 8.4 and 8.5 above is that probabilities, especially personal probabilities, are not precise quantities, and that it therefore makes little or no sense to represent them by real numbers, especially irrational numbers. This brings me to a second application of ideas from chaos to the criticism of Bayesianism; here too, I think, more mainline objections are available, and these new ideas serve only to intensify the attack. The issue is that of indeterminate probabilities and approximate distributions. Again we may refer to Howson and Urbach, who acknowledge that it is often not possible to do more than specify imprecise bounds for the odds we think fair, and "there will be an intermediate interval of odds between which we are unable to discriminate and whose precise extent, if indeed it has a determinate length at all, reflects the extent of our ignorance" (*op. cit.*, 69f). Howson and Urbach are not exercised by this problem; they think that "due allowance can be made, if necessary, for any imprecision in the prior distributions" (*loc. cit.*). It seems to me that this opinion is decisively contradicted, not only by the speculative and controversial ideas of 8.4 and 8.5, but by the banal ideas of 8.3. Referring back to the example there discussed, let us suppose that an agent can place $P(\text{Heads})$ only in the interval $(0.5, 0.502)$. What probability ought he to assign to the event X? It is not quite clear what Bayesianism's answer to this question is, unless $P(\text{Heads})$ is regarded as a random variable with its own probability distribution. But although it can, of course, be so treated, such a treatment would make necessary new proofs of the laws of large numbers. (As I understand the matter, this resembles the approach to indeterminate and second-order probabilities of Levi 1974 and of Gärdenfors and Sahlin 1982; 1983. I cannot say anything about these developments here.)

The comparison that Howson and Urbach make between the looseness of probability assignments and the imprecision of physical magnitudes is, in my view, simplistic. It may not make much sense, because of the existence of molecular fluctuations, to talk of the precise value of the length of a rod, but we may be able to talk sensibly of its precise mean value (over time), even if we cannot measure it exactly. There is more or less a precise value around, but we are a bit vague about it. But this is not how it is in the probabilistic case. If the value of $P(X)$ is not precise, it will be smeared over much of the unit interval, whatever the approximate value of P(Heads). So unless the agent is almost completely precise about X, he cannot, even as an approximation, be thought of as assigning any definite probability in the interval (0.5, 0.502) to the event Heads. I am aware that much more could be, and has been, said on the question of interval-valued probability. My point here is to draw the theory of chaos firmly into the picture, and to make as plain as I can how very difficult it may be to defend the view that approximate distributions are good enough.

In short, the objection that probabilities are not precise rebounds on those who press it. If probabilities are not precise, they are nothing.

8.8. Conclusion

It was long thought that dynamical systems governed by the equations of classical mechanics are typically stable or periodic. Those systems (such as the atmosphere, and other turbulent fluids) that resisted all such analysis were brushed aside as too complicated. We know better now. Strange attractors are as typical as point attractors and limit cycles. Their mathematics may be harder, but there are plenty of them both in principle and, it seems, in practice. Might the same be true in the theory of probability? We are all familiar with parametric distributions (such as the normal, the binomial, and the Poisson distributions) that are very decently behaved in the face of mountains of data, whether handled by Bayesian methods or by those of classical statistics. Has their familiarity blinded us to the possibility of distributions that are less well behaved? If not that, has it led us to think that such distributions must be outliers, of zero significance? I have little doubt that it has so persuaded us, and that the conclusion is no more defensible than it was in the case of classical mechanics. Just as the stability of dynamical systems in general was an explanandum that never found its explanans (because it is false), so the objectivity that Howson and Urbach (for example) see almost everywhere in scientific activity and are at pains to explain may be exceptional, not typical. Or, if I may suggest a less extreme position, it may be that distributions of the kind that I have presented are exceptional in science, but more typical in everyday life. I have no idea whether this is the case or not. If it is, it might provide a germ of a Bayesian explanation of why there is somewhat more agreement in science than there is outside it.

It must however be stressed that many questions remain open; not only the chaotic character of the distributions themselves, which I have made no proper attempt to establish, but especially the question of whether ill-behaved probability functions of the kind illustrated above are typical (in the space of possibilities). You may well feel sceptical about this, noting the artificiality of the constructions. The artificiality is cheerfully conceded. But I would remind you that from the viewpoint of real dynamics the logistic function too is not much more than a toy. It is an educational toy because it forces us to get clearer about the circumstances under which stable behaviour is to be expected. I hope that the chaotic distributions of 8.4 and 8.5 above, even if they turn out to be genuinely atypical, may in the same way encourage those who hold that convergence of opinion is a hallmark of objectivity to spell out more plainly the conditions under which such convergence will be found to obtain.

Technical Note

The following are the data streams illustrated in Figures 5–8 above.

Figure 5:   111111010101101011110001100010011101001
1011010111100101110111111011011010000111000010010
1000011110101011110110111111111100111111001101010
0000101000011001100111111001010010011100111101011
1000011000000101000101100001101010010010011000110
0100010001010110111101100010011001001101010010010
0001111111000111101101110101100011110100100101100
0101010011100101000111110000010101111011100000010
1101100000111010101001000

Figure 6: 0100111111110000011001101110110110100001
1001011011101001011011111011111001011101100011 11
0101001000000111110110111110011111010100101 11101
1001000010111000001100110101010101001011011 01001
0100110100011100001100010110010101100101100 11000
0100111011110101100110110001000101001010000010111
1011001001010000001100010010111011011 01000111010
1111010100110100110000100100111000110110101 11110
10000010010011001111 0011

Figure 7: 1011010110100101011010111111111100001100
0110110101000101001110010101001001100101001 10111
1011101011111010010010010011010001111001111 01110
0110001100000100110110101100011101111100001 01110
0001101010100001010111111011110100011000101 01011
0111111001010100001110111011000011111100100 10001
00010010000100010110111101101001100111111111 1110
0010111001001011001011010001101110100101111 10100
00101100110100101001 0000

Figure 8: 0110001001111000011001001011101100110111
0111011110000001001010101110001101010010110 1110
1010111111100111001111100100100100100 10000010111
00001101011001011110001011011101101100010011 1111
0010110011100111000111010000001110011110011 011
1001000100111001010010101101111001011100001 10001
0100000010101100001110100001001001100011111 10111
001110110011011001000010010010001111001101 11111
1010111010111111110000110

9
Objective Probabilities

9.1. Introduction

One of the principal challenges confronting any objectivist theory of scientific knowledge is to provide a satisfactory understanding of physical probabilities. The earliest ideas here, known collectively as the frequency interpretation of probability, have now been all but abandoned, and have been replaced by an equally diffuse set of proposals all calling themselves the propensity interpretation of probability. The principle virtue of the propensity interpretation in any of its variants is supposed to be that, unlike the frequency theory, it renders comprehensible single-case probabilities as well as of probabilities in ensembles and in the long run. 9.2 recalls salient points of the frequency theory of von Mises (1928/1957), and 9.3 looks at some of the improvements introduced by Popper in Chapter VIII of his 1934/1959a, especially those that point towards later ideas. In 9.4 I filter out from Popper's numerous expositions of the propensity interpretation what I consider to be its most interesting and fertile strain. I then go on to assess it. First I defend it, in 9.5, against recent criticisms to the effect that conditional, or relative, probabilities, unlike absolute probabilities, can only rarely be made sense of as propensities. Finally in 9.6 I shall make an effort to challenge its predominance by outlining a rival theory: an irreproachably objectivist theory of probability, fully applicable to the single case, that interprets physical probabilities as instantaneous frequencies.

9.2. The Frequency Interpretation of Probability

It is pleasant to be able to begin by advertising one of the under-rated virtues of the frequency interpretation of probability: that (in common with subjectivist interpretations) it makes sense of ascriptions of probability in a world that obeys strict causal (or deterministic) laws. In such a world there are no uncaused individual events or effects. But there may be mass or statistical effects (such as equilibrium states) that are neither in conflict with the causal laws nor deducible from them. As von Mises emphasized, the frequency interpretation makes possible an objectivist construe of the probabilities that occur in classical statistical mechanics (1928/1957, 181–83; see also Popper 1934/1959a, Section 70). According to von Mises, probabilities belong not to individual events, but to kinds of events (attributes) in an unending repetition of an experiment or natural recurring phenomenon. In dicing, for example, typical attributes are *six-up* and *even-number-up*; their probabilities are not determinable a priori (as in the classical theory) nor a reflection of our expectations (as in the many allotropes of the subjectivist theory). On the contrary, $p(six-up)$ is an (idealized) physical quantity: the limit to which the relative frequency (Number of sixes)/(Number of throws) will tend as the die is thrown and thrown again. A collective (*op. cit.*, 28f) is a potentially infinite sequence in which these limiting frequencies exist (the axiom of convergence) and remain unaltered in any subsequence selected by a disinterestedly specified gambling system (the axiom of randomness); that is, any subsequence of the original sequence that is selected without regard to the attribute actually instantiated. It is well known that 'any subsequence' here is much too strong. There are inevitably many selections that give quite different results. The idea, however, which is clear enough, is that it should be possible to sift out deviant subsequences according to some method; hence von Mises's alternative description of the axiom of randomness as the axiom of excluded gambling systems. In the frequency theory the probability of an attribute is defined as the limiting value of its relative frequency, and is defined only relative to a collective (sometimes referred to as a reference class).

It is not excluded, therefore, that the same attribute should have different probabilities in different collectives. A pet example of von Mises's own was that of the life tables of insurance companies (*op. cit.*, 16–18). The attribute *death before age 50* is found empirically to have different long-run frequencies in the class of men aged 49 and in the larger class of men and women aged 49. If these classes are authentic collectives, the probability of *death before age 50* varies with the collective to which it is referred; and there is doubtless a multitude of others in which the attribute has some other probability. It makes no sense, von Mises concluded, to talk of the probability of a single event, such as the death before his 50th birthday of a particular man aged 49. Popper agreed (*op. cit.*, Section 71), acknowledging that any singular statement of probability makes implicit reference to "the objective statistical state of affairs" (and hence to a collective).

Susceptibility to death seems, however, to be a very individual trait, one for which intuitively we feel somewhat inclined to ascribe probabilities (known or unknown) to single cases. There is a problem here for the frequency theory. One way out might be to deny that the class of 49-year-old men and women is a genuine collective, on the grounds that there is a way of selecting members of the class (sexual selection) that affects the long-run frequency of the attribute *death before age 50*. In other words, we might define collectives as sequences in which relative frequencies are insensitive not only to von Mises's place selections (or Popper's neighbourhood selections) but to gambling systems specified in an empirical way. But this move would annihilate the frequency theory as von Mises conceived it. For as he himself observed (*loc. cit.*), by taking into consideration more and more of an insurant's properties we shall end up not with a collective but with a set containing the insurant alone; a set, that is, in which all frequencies are 0 or 1. The same is true of games of chance and of natural random processes if the time of occurrence is allowed to serve as a method of selection. And even if the collective is not drastically impoverished, we cannot in a deterministic world fully avoid the risk of trivializing the probability distribution. There may well exist, for instance, some specifiable conditions that guarantee the occurrence of *six-up* with a die. Should we select throws of the die in accordance with these conditions, the probability of the attribute *six-up* in the resulting collective would necessarily equal 1.

It is only by restricting probability assignments to mass effects that the frequency theory manages successfully to reconcile statistical phenomena and deterministic laws. The price it pays for this is to leave single-case probabilities uninterpreted. Subjectivism also is compatible with determinism, and makes fair sense of single-case probabilities; the price paid by this approach is the divorce of statements of probability from objective statistics. In 9.4 and 9.5 I consider the propensity interpretation, which is an objectivist interpretation where single-case probabilities are supreme, and determinism is surrendered, and in 9.6 I tentatively suggest an interpretation, free of subjectivist taint, in which single-case probabilities, mass phenomena, and both indeterministic and deterministic laws are able to co-exist. But before then I want to look at some other problems that plague von Mises's version of the frequency interpretation, and to see how they may be quelled.

9.3.   The Falsifiability of Probability Statements

The two main problems tackled in the probability chapter in *Logik der Forschung* (1934/1959a) were what Popper called "the fundamental problem of the theory of chance", which is placed on the agenda in Section 49, and "the problem of decidability", which, though briefly adverted to earlier in the book, is finally brought to heel in the closing sections of the chapter (beginning with Section 65). Though formulated in the context of the frequency interpretation of probability, each of these problems is of incontestable significance for any properly objective physical interpretation. What Popper calls the fundamental problem is the problem of explaining the occurrence of order in the midst of disorder. How is it that an apparently lawless sequence, such as a sequence of tosses of a coin, a sequence that seems utterly resistant to detailed prediction, can nevertheless evince such regularity in the large? In short, what is the source of the statistical stability of random sequences? If they are truly anarchal, why do their relative frequencies not jump around equally wildly? The problem of decidability—better called the problem of falsifiability, since it is manifest that no statement of limiting frequency can be verified—hinges on Bernoulli's theorem, or some other law of large numbers. There can be no doubt that every combinatorially possible finite sequence of outcomes (of tosses with a coin, say) has a probability greater than zero. On a crude frequency reading it follows not only that every arrangement of outcomes is physically possible, but that it will actually happen. Thus no possible sequence is ruled out by any probabilistic hypothesis. In other words, statistical hypotheses are empirically unfalsifiable, and so by Popper's criterion of demarcation they are to be classed as unscientific.

   In a sense, these two problems are different faces of the same coin: the more serious we find one of them, the less serious we shall find the other. On the one side, what Popper calls the fundamental problem is not a significant one if what we observe as statistical stability is simply an appearance generated by our incapacity to survey sequences of indeterminate length; if no legitimate order actually does arise out of the welter of disorder, then we need not spend our time trying to explain it. But if we take to heart this claim that observations of finite stretches of random sequences are uninformative, the problem of falsifiability becomes acute. On the other side, if we allow that the behaviour of a finite initial segment of an infinite sequence may well be able to tell us something about the whole sequence, so taking the sting out of the problem of falsifiability, then we shall have to face the full force of the problem of stability. To put the matter briefly: our assessment of the relative intransigence of the two problems at issue will depend crucially on how faithfully we hold to the frequentist doctrine that it is the limiting relative frequencies in a collective, and only these limits, that may be identified, even approximately, with the collective's probability distribution. To a reactionary frequentist, stability is no more than a trivial consequence of the existence of a limiting relative frequency. To a radical frequentist it is the triteness of this response that renders it useless as a solution of the fundamental problem.

In *Logik der Forschung* Popper was a frequentist filled with reforming zeal, and this zeal was channelled directly into a new solution to the problem of stability. What he did was to show how von Mises's axiom of convergence, which explicitly postulated that relative frequencies converged to a limit, could be dropped without rendering the frequency theory impotent (1934/1959a, Section 64). Later Popper's reformism burgeoned into a fervour so revolutionary that the frequency interpretation was cast out altogether, and replaced by the propensity interpretation, which may be seen not only as an attempt to solve the problem of the single case but also to provide a deeper solution to the problem of stability. Indeed, in his discussion of the argument known as 'Landé's blade', an argument designed to show that no deterministic solution to the problem of stability does more than initiate an infinite regress, Popper explicitly adduces the need for a single-case interpretation of physical probabilities (1982a, Sections 29f). Unquestionably we need them for many practical applications. But that is not important here. What is crucial is that probabilities that apply to the single case are neither frequencies nor, though propensities, propensities to produce frequencies. They may therefore have the ability to explain why it is that frequencies converge, why random sequences are statistically stable.

But this is to anticipate. In *Logik der Forschung* itself Popper undertook to solve the fundamental problem of the theory of chance by discarding the axiom of convergence altogether and strengthening the axiom of randomness to such an extent that the existence of limiting frequencies could be derived from it. This was done by requiring not just that selected infinite subsequences of a collective be statistically similar to the whole collective, but that finite initial segments be so too. Popper gave an ingenious, but nonetheless extremely simple and intuitive, definition for what it is for a finite sequence to be random, and proposed that probabilities be restricted to sequences that, by the lights of this standard, are ideally random right from the beginning. It was then demonstrated that relative frequencies necessarily converge when such ideally random sequences are extended indefinitely far. Note that, in addition to giving in this way a solution to the fundamental problem of the theory of chance, by showing that statistical stability is a consequence of local disorder rather than in conflict with it, Popper's proposal also gives a solution to the problem of falsifiability. For, whatever a sequence's distribution may be, its claim to be random, and thus to be a licensed vehicle for the carrying of probabilities, is simply refuted if its initial behaviour (not just its eventual behaviour) is sufficiently remote from the ideal. Random is as random does. Reading Popper's words in the strongest possible way, we may say that when a fair coin is tossed there are some sequences of heads and tails that on their own are genuinely physically impossible. This is not to say that their probabilities are zero, which clearly they are not; these unrepresentative sequences can even occur, but only within the context of more commodious, better regulated, sequences. Probabilistic hypotheses in this way become strictly falsifiable by empirical means; something that everyone really knew all along that they must be, though no one knew how they were.

It should be noted that Popper's definition of finite random sequences was criticized quite early by Ville (1939), and was not fully refined until later; see Popper 1959a, Notes *1 and *2 added to Appendix iv. As far as I know, Popper was the first to consider the possibility of randomness in the finite case. The idea later came to prominence in the computational-complexity approach to randomness developed by Kolmogorov, Martin-Löf, Chaitin, and others. For an excellent survey of this area see Delahaye 1993.

There are two obvious, and related, objections that can be raised against this proposal. The first is that the fate of a probabilistic hypothesis seems to be unduly sensitive to the moment at which we start to test it. Is an old roulette wheel that turns up 0 ten times in successive games to be regarded indulgently as a mere victim of statistical circumstance, whereas a brand new wheel that does the same is at once returned to the manufacturers as unbalanced? For after all, even in an ideally random sequence a string of ten successive 0s will occur within the first 37^{10} games (about 4.8×10^{15} games, presumably enough to grind most real roulette wheels to dust). The second objection is that some deviation from the demands of ideal randomness will of course be tolerated in any physically realized sequence; we shall not reject a coin's claim to fairness just because its first two tosses yield heads (even though no ideally random sequence with equidistribution can start with two heads or two tails). It is tempting to think that these objections are really the old problem of falsifiability in another guise, and that the digression into finite random sequences has made no genuine impact on the problem. This, it seems to me, is not so. In the first place it neglects the fact that, for better or for worse, the frequency theory really does ascribe probabilities to attributes within sequences rather than to devices or chance sets-up. If the sequence we are concerned with is the sequence of all tosses with a particular coin, then our main problem in performing a statistical test is that we rarely, if ever, know exactly where in the sequence we are. What we do know, however, is that representative subsequences (those whose frequency distribution approximates well to that of the whole sequence) are to be found almost everywhere in the sequence, whilst grossly deviant subsequences turn up only a long way, usually an exceptionally long way, from the start. In most cases we know also that only a relatively small number of trials have been performed on the set-up. A robust test of a probabilistic hypothesis, in consequence, one, that is, that is not painfully sensitive to what assumption we make about our location, will be a test that, in effect, assumes that we are at or near the beginning of the sequence under consideration. It would be intolerable that a probabilistic hypothesis should be guarded from falsification by directing the blame instead at an auxiliary hypothesis that was (though essential to the test) itself scarcely susceptible of independent test. This insistence on robustness, with its refusal to provide round-the-clock protection to probabilistic hypotheses, is a kind of dual to Popper's requirement, for statistical and non-statistical hypotheses alike, that we count as falsifications only those effects that are reproducible. Indeed, we should "take the methodological decision never to explain physical effects, i.e. reproducible regularities, as accumulations of accidents" (1934/1959a, Section 68) in just the same way that we take the decision never to explain accumulations of accidents, i.e. non-reproducible irregularities, as physical effects (*op. cit.*, Section 22). The first decision stops falsification from becoming hopelessly difficult, whereas the second was introduced to stop it from being absurdly easy. As Popper himself stresses, these considerations apply to deterministic and probabilistic hypotheses indiscriminately. The problem of falsifiability is in this way, it seems to me, solved in full accordance with the methodological principles enunciated earlier in *Logik der Forschung*. Probabilistic hypotheses, even when interpreted in the frequency manner, can be rescued from the charge of being "impervious to strict falsification" (as Popper puts it in his introductory remarks to Chapter VIII of his 1934/1959a).

9.4. The Propensity Interpretation of Probability

"The problem of the single case *is* the problem of the reference class" writes one frequentist (Salmon 1967, 91). The device employed by the propensity interpretation to eliminate the problem of the reference class (that is, the collective to which ascriptions of probability must be referred), and hence solve the problem of the single case, is simply this: to eliminate the reference class. In the propensity interpretation, the probability of an outcome is not a measure of any frequency, but (as will be explained) a measure of the inclination of the current state of affairs to realize that outcome. In not treating probability as anything to do with a frequency—but scarcely in any other way—the propensity interpretation resembles subjectivistic interpretations, in which probabilities are understood as degrees of belief or knowledge. It was argued in Chapter 7 that something akin to the problem of the reference class still arises for subjectivism, in the problem of which body of evidence probabilities should be evaluated in relation to.

Popper's first presentation of the propensity interpretation (1957/1983) was a terse prospectus of some of the arguments of *The Postscript*, each volume of which takes up some aspect of the new interpretation (amongst other topics). Part II of Volume I (1983) assails subjectivist interpretations of probability, points to some insufficiencies of the frequentist view, and introduces the idea of probabilistic propensities as generalized forces. In Chapter IV of Volume II (1982a) they are indissolubly interlocked with indeterminism. In Volume III (1982b) the propensity interpretation is applied to the resolution of paradoxes in quantum theory. There have been several other expositions (1959b; 1967a/1982b; 1974b; 1989; 1990). But almost the clearest statement Popper has given of what is at issue is perhaps to be found at a place where he repudiates such an interpretation of probability. In an early discussion of the subjectivist interpretation he wrote (1934/1959a, 212):

> I do not object to the subjective interpretation of probability statements about single events . . . so long as we clearly recognize that the *objective frequency statements are fundamental.* . . . I reject . . . any interpretation of these formally singular probability statements . . . as statements about an *objective state of affairs,* other than the objective statistical state of affairs. What I have in mind is the view that a statement about the probability $\frac{1}{6}$ in dicing is not a mere confession that we know nothing definite (subjective theory), but rather an assertion about the next throw—an assertion that its result is objectively both indeterminate and undetermined—something which as yet hangs in the balance.

A footnote appended to this passage in 1959 makes it quite plain that the position here rejected is at the heart of the propensity interpretation. Singular probability statements are not, on this interpretation, covert statistical statements, but statements of the propensity of the balanced —but perhaps not evenly balanced—state of affairs to eventuate in one outcome rather than another; statements of incomplete or generalized causation. No frequency lurks behind the probability, and so no reference class is needed for the probability to be referred to. The probability of an outcome is no longer conditional, or relative, in the way the velocity of a body is—that is, undefined in the absence of a reference frame, and capable, for different frames, of having different values at the same time. It is conditional, or relative, only in that it is a function of the conditions, and may well assume different values at different times, or under different conditions. But at any one time only one (complete) set of conditions—state of affairs—obtains.

According to the propensity interpretation, in the strong form in which it is recommended here, what we call *the* probability of an outcome is as absolute (though not as inexorably impermanent) a characteristic as is the age of a person: it is relative only to the unique situation of the world (or the causally operative part of the world) at the time. A good example of a varying propensity is life expectancy, whose unsatisfactory handling by frequentism was frowned on above. The probability that I survive one more year from today varies day by day, being affected by my activities (and others'), by the progress of medicine, and ultimately—if I live long enough—by inescapable facts of human biology (Popper 1990, 14f). None of this should be allowed to suggest that the propensity interpretation excludes talk of conditional probabilities. The propensity (today) of my survival for another year, given that I take up free-fall parachuting tomorrow, is, I imagine, a bit different from the propensity (today) of the same event, given that I devote myself to full-time raffia work. Each of these propensities generates a conditional probability $p(a|c)$, which is best read subjunctively (as it would be in the subjectivist interpretation): it is what the propensity of a would be were c to obtain. (In 9.5 I say more about this.) But neither of these conditional probabilities, nor any of the countless others that there are, is what we normally mean by *the* probability (today) that I survive; this depends (in a perfectly familiar way) on the absolute probabilities (today) that the various conditions mentioned (and unmentioned) themselves come to pass. Note that these conditional probabilities are as much single-case probabilities as the absolute probability is; they are concerned with *my* survival for one year from *today,* with the falling six-uppermost of *this* die on its *next* throw, and so on.

Understood as propensities, that is, conditional probabilities are conditionalized twice over. Absolute probabilities are relative to the prevailing conditions, to the total present state of the universe (the infinite regress here is not crucially different from the regress that arises in classical mechanics if we presume to run a Laplacean universe backwards from its present state), and true conditional probabilities are relative to further as yet unrealized conditions. Note that, by reason of the first relativization alone, a calculus of conditional or relative probabilities is needed if we are to do full justice to the propensity interpretation (see Popper 1974b, 1131–37; 1990, 16), rather than a system (such as that of Kolmogorov 1933/1950) in which probability is represented by a one-place functor. The latter kind of system—in which $p(a|c)$ is defined by the quotient $p(a\text{-}\&\text{-}c)/p(c)$—suffices only if 0 propensity is construed as no propensity, emphatic impossibility. Popper suggests this reading more than once (1990, 13, 25; for a slightly weaker position see 1983, Part II, Section 24), even though it renders truly miraculous something that seems to happen time and time again, namely the fact that "new possibilities are created, possibilities that previously simply did not exist" (Popper 1990, 19). This is a serious matter, but I cannot go further into it here.

As already observed, in the propensity interpretation there is no frequency lying behind the probability. On the contrary, the probability or the propensity lies behind, and is required to explain, the observed frequency. It is only if the propensity of an outcome remains constant in globally different (though locally similar) situations that we can expect to measure its true value empirically; we shall need, that is, either an ensemble (such as a swarm of like-minded particles) or a repeatable experiment (such as dicing in a comparatively stable milieu). It is the sameness (which, of course, we can know only by guesswork) of the propensity or probability in different situations that makes the situations similar, and not (as in the frequency interpretation) the other way round. Popper says all this pretty clearly: ". . . if we wish to *test* a probability statement, we have to test an experimental sequence. But now the probability statement is not a statement *about* this sequence: it is a statement *about* certain properties of the experimental conditions" (1957/1983, 68). It is to be regretted, therefore, that alongside such trenchantly opposed presentments of the propensity interpretation and its elder brother we find remarks that describe the new proposal as "a revised or reformed statistical interpretation" (1957/1983, 65), or depict propensities as "tendencies to produce relative frequencies on repetition of similar conditions or circumstances" (1982b, 71). To be sure, tendencies to produce frequencies exist in certain cases, or they seem to. But they are highly derivative, and the fundamental propensities from which they derive might still exist even if there were—perhaps because of failure of various necessary independence conditions—no recognizably stable frequencies anywhere, at any time. The propensity interpretation, I do emphasize, is a genuine alternative to the frequency interpretation. It grew out of it, yes; but it has since grown out of it.

Popper's remark that propensity statements are statements about properties of the experimental conditions must not be misunderstood. The assertion that the probability of *six-up* is a property of the die-throwing apparatus (and nearby surroundings), though not incorrect, does not mean that the probability of *six-up* measures the strength of a connection between the outcome *six-up* and an incompletely specified situation (hence, in effect, a class of similar situations). It is an empirical assertion that happens to be true but might easily have been false. (Were it false, hypotheses about the probability of *six-up* would be a great deal harder to test.) Indeed, in the case of life expectancy the parallel assertion—that the probability of my death within a year is a property of my local surroundings—is mistaken; the present work, or enthusiasm for work, of distant medical researchers is an important determinant too. Propensities are not located in physical things, and not in local situations either. Strictly, every propensity (absolute or conditional) must be referred to the complete situation of the universe (or the light-cone) at the time. Propensities depend on the situation today, not on other situations, however similar. Only in this way do we attain the specificity required to resolve the problem of the single case.

From a formal point of view the main difference between the two interpretations is that they are based on different probability spaces: the frequency interpretation is based on a space of events, incompletely specified situations, present and future, such as throws of a die; while in the propensity interpretation the probability space is the set of all future, completely specified, possibilities that might disembogue from the present. (Thus the probability space is itself perpetually changing.) On this more abstract and metaphysical space we can, if we want to, define probabilities as proportion (or measures). Then the probability today of my surviving another year is the proportion of possible states of the world in one year's time in which I am still alive. For there to exist probabilities other than 0 and 1 it is therefore essential that the class of future possibilities contains more than a single element. Though in substance very different, such a definition of probability is formally identical with the propensity interpretation, and it reveals very clearly that only in a world that is genuinely indeterministic—not necessarily in all aspects, but at least in part—is the propensity interpretation a proper interpretation of probability. (I therefore cannot agree with Popper that "[t]he propensity interpretation is . . . that of classical statistical mechanics" [1983, 398], though I do agree that it is the correct interpretation to apply to the statistics of gases.) It goes beyond the bounds of this chapter to address the question of whether the requisite indeterministic loop-holes really exist in games of chance, mortality tables, and other macroscopic phenomena to which probabilistic ideas are customarily applied (see Giere 1963).

9.5. Criticisms of the Propensity Interpretation

The propensity interpretation has prospered mightily in the last 30 years, and in one version or another, perhaps more muted than the one described here, it is widely accepted as an important component of our understanding of physical probability. There are inevitably differences of emphasis, and of terminology: Suppes, to take one example, writes that "propensities are not probabilities, but provide the ingredients out of which probabilities are constructed" (1987, 345). Subjectivism, to be sure, is still much favoured, despite years of chronic underperformance (especially when it comes to the problem of explaining statistical stability), but there are several Bayesians (for example, Jeffrey 1977, Lewis 1980) who recognize the need for an objectivist and non-frequentist theory of probability. Lavish criticism has accompanied appreciation, of course. Among the commonest criticisms of the propensity interpretation are the complaint that—because there are no propensities of any sort—it is inapplicable in the only kind of world entertained by traditional empiricism (Mackie 1973, 179–187), and the complaint that—because there are no non-trivial propensities—it is inapplicable in the only kind of world entertained by traditional determinism (Howson and Urbach 1989, 222–25). These objections were anticipated by Popper in the *Postscript*—especially in the Metaphysical Epilogue in 1982b and in Section 27 of 1982a. I shall not here consider either matter further. It is often maintained too that propensities are properly lodged not in abstract situations or states of the world, but in concrete physical objects such as dice, die-throwing machinery, and persons (Mellor 1971). This point too has been thoroughly dealt with in Popper 1983, 359f, and has been adverted to above.

In this section I intend to recount briefly, and to answer fully, what is, to my mind, the most interesting line of criticism that the propensity interpretation has yet attracted. At first sight the criticism appears unanswerable, for it appears to pick out with awesome clarity a formidable obstacle to the interpretation of conditional probabilities as generalized causes. Nonetheless, I think that the criticism is mistaken, and that it betrays a deep misunderstanding of the theory that it takes to task. The objection, very shortly stated, is that although there are many choices of a and c for which the conditional probability $p(a|c)$ can readily be interpreted as a measure of a propensity or generalized cause, there are others for which such an interpretation is nonsensical. The reason is that causes have a characteristic direction, and—to put it figuratively—sometimes a is not correctly aligned in relation to c. For example, if c is the set of conditions under which a die is thrown, and a is *six-up*, then $p(a|c)$ is effortlessly interpreted as a propensity. But $p(c|a)$ makes not much sense; the outcome *six-up* has no influence on the conditions of the experiment that produced it, and it has no special propensity to produce them. Likewise *death before age 50* has no special propensity to make a man aged 49 take up free-fall parachuting. If we could set $p(c|a)$ equal to $p(c)$ in all such cases, there would indeed be little to get excited about. But it is not hard to construct examples in which such an assignment would lead to a violation of Bayes's theorem—which says that $p(c|a)$ is identical with $p(a|c)p(c)/p(a)$. This criticism has been pointedly made by Humphreys (1985, 559–563). Nor is the difficulty restricted to those examples in which c is a set of generating conditions for a, or temporally anterior to a. Putting a decent construction on the term p (*six-up* | *even-number-up*) in the case of a fair die is no less awkward; for the appearance of an even number must be at the same time the appearance either of a 2, or of a 4, or of a 6. No relevant feature of the outcome remains undetermined, and hence there is no noticeable room for an interpretation of the probability value ⅓ as an indeterministic propensity. This objection is due to Milne 1986, 130f.

One solution—a dismal and defeatist one that I shall not pursue—is to drop the intuitive requirement that $p(c|a)$ be defined, or interpreted, whenever $p(a|c)$ is. In addition to its other demerits, this dodge would be an uncomfortable companion for one of the desiderata that Popper imposes on a formal theory of probability (1959a, 326f). Another unpromising method of resolution is surrender: the bland announcement that, after all, the propensity interpretation must be a non-standard interpretation of probability theory, obeying some but not all its laws (in the same way as Riemannian geometry is not a standard interpretation of Euclidean geometry). What lies behind this suggestion, due to Fetzer 1981, 285, is no more than a distinction between direct and inverse probabilities. Fetzer does not explain whether a term like $p(c|c)$, which features in one of the central axioms of the theory of probability, is to be taken to represent a direct probability or an inverse probability. Indeed, we might well wonder whether any state of affairs has a propensity (even of maximal strength) to produce itself.

I have tried to state the objection convincingly, even though I am now persuaded that it is founded on a serious misapprehension of the status of conditional probabilities in the propensity interpretation. In fairness I should say that this misapprehension is not, as far as I am aware, plainly contradicted in any of the standard presentations, and may even be encouraged by the unqualified characterization of probabilities as properties of experimental arrangements. As noted in the penultimate paragraph of 9.4, propensities are not properties of incompletely specified situations, but of the completely specified present situation. What we call *the* probability of *six-up* is relative to the tightest possible specification of the present and the loosest possible specification of the future. It is today's probability, conditional on nothing at all (the tautological condition), and accordingly an absolute probability, that the present unique situation should develop into one in which *six-up* comes about. Conditional probabilities are conditional, as is usual, on subsets of the probability space, the space of today's future possibilities. Notice that the set of events that constitute throws with a particular die is not even a subset of this probability space, so it is starkly impossible that any conditional propensity should be conditional on such a set. All propensities, absolute or conditional, are simply propensities of the present unique situation to develop into another.

Thus if a is my survival one year from today, and c is my taking up parachuting tomorrow, $p(a)$ is the propensity for today's world to develop in a year's time into a world in which I am still alive. That is straightforward. Similarly the conditional probability $p(a|c)$ is the propensity for today's world to develop into a world in which I am still alive in one year's time, given that it—today's world—is a world that develops into one in which I take up parachuting tomorrow. So far, so good. Note that the causal influence that is measured by $p(a|c)$ is an influence from today to a day one year hence (and, indeed, since death is irreversible, to all intermediate times). It is not an influence from the time recorded in c to the time recorded in a. But in the case that c temporally precedes a, there is plainly no harm in giving $p(a|c)$ the standard reading as the propensity of c to produce a.

What about the inverse conditional probability $p(c|a)$? This comes out as the propensity for today's world to develop into a world in which I take up parachuting tomorrow, given that it—today's world—will by the end of the year have developed into one of those worlds in which I am still alive. It is not suggested that my later survival has any direct causal effect on my earlier going in for parachuting. But it does not follow that $p(c|a)$, so interpreted, must be the same as $p(c)$. The causal pressure is from today to tomorrow, not from the remote future to tomorrow. It is perhaps easiest to think of $p(c|a)$ subjunctively as what the propensity for me to indulge in parachuting tomorrow would be today if something were to happen today that guaranteed my survival for another year. In the same way, $p(six\text{-}up|even\text{-}number\text{-}up)$ is the propensity for today's world to become a world in which six is thrown, given that it is going to become one in which an even number is thrown. Such propensities have no direct operational meaning, it may be conceded. That is often the way with the subjunctive mood.

There is no need to conclude that only absolute probabilities are interpretable as generalized indeterministic causes. Nor need we resign ourselves to reading $p(a|c)$ as no more than the ratio of the strengths of two unconditional or absolute propensities (Sapire 1991, 338–40). The relative probability $p(a|c)$ is a genuine propensity, but it is a propensity, conditional on *c*, of the contemporary state of the world, not a propensity of the condition *c*.

I cannot leave this business without saying that, in my view, the loss of conditional propensities—were it necessary—would not be as devastating to the propensity interpretation as Milne, for example, seems to think. The ground for his gloom, that "as Kolmogorov pointed out, what distinguishes probability theory from measure theory is the definition of conditional probability" (*op.cit.*, 131), is, I am afraid, a simple misunderstanding; and it would after all be most strange if an eliminable definition were at the heart of probability theory's individuality. Nor could it be the more general calculus of conditional probabilities that Kolmogorov had in mind, since his own axiomatization is notable for its reduction of conditional probabilities to absolute probabilities. What Kolmogorov actually says is: "The concept of mutual *independence* . . . holds, in a certain sense, a central position in the theory of probability. . . . We thus see, in the concept of independence, at least the germ of the peculiar type of problem in probability theory" (1933/1950, 8). What this must mean, quite plainly, is that it is the postulation of independence that marks the difference between probability theory and measure theory. The laws of large numbers, for example, which are not usefully formulated in terms of conditional probability, rest on hypotheses of independence (or something similar). Theorems of measure theory typically do not.

9.6. Probability as Instantaneous Frequency

The frequency theory of von Mises is not a local theory (to adopt unceremoniously some meta-quantum-mechanical jargon): the probability that a man aged 49 will survive another year depends not just on that man, but on all other men aged 49; the probability that a die falls six when thrown in a prescribed way is a function of all throws made in that way. The problem of the single case is not that we cannot define single-case probabilities, for we can define too many of them. It is that, to attain uniqueness in anything like a uniform manner, we seem to have to pick a probability that is altogether local, and thus altogether trivial. It is the chief merit of the propensity interpretation that it eludes this difficulty: *the* probability of an outcome at a time is specified uniquely by relativizing it to the total world situation at that time. In an unmistakable sense, that is to say, propensities are not locally defined either, although, as we have noted, they may perchance not be affected by distant events. Radioactive decay is perhaps the best known example of a propensity that is largely insensitive to external influence. But there is a difference between being locally determined, which is a factual matter, and being locally defined, which is not. My tax inspector annually reminds me that my annual income is not locally defined, but includes anything I earn overseas; but in some years, as it happens, my income is locally determined.

I say this not in criticism of the propensity interpretation, but in order to prepare the ground for the new interpretation of probability that I shall sketch in this section. For just as Popper, in endorsing the propensity interpretation, pays tribute to its predecessor, emphasizing that "a frequency theory can be constructed which is free of all the old difficulties" (1959a, 362; see also 1983, 347f), so too do I wish to acknowledge the strength and the achievements of the propensity interpretation. The virtues of the interpretation to be outlined are that it is local, and that—unlike the propensity interpretation—it makes good sense of probability in outposts of the world that are tyrranized by deterministic laws. I am not a cent per cent determinist any more than Popper is, but current work in deterministic chaos makes it undeniable that the universe may be rather more deterministic than had previously been thought. It may be useful to have at our disposal an interpretation of probability that can be accommodated in these altered circumstances.

In 9.4 I tried to represent the propensity interpretation as a remarkable leap forward, rather than a mere refinement of the frequency interpretation of von Mises and Popper. This was despite some of Popper's own evaluations—in particular those in which he described propensities as dispositions to produce frequencies (1957/1983, 67). That physical probabilities are properties of parts of the world ("like mass, specific heat, or electrical resistance") is an idea we find already in von Mises (*op. cit.*, 14). Their dispositional character is sometimes asserted also by him, for instance when he says that "a collective is . . . a long sequence of observations for which . . . the relative frequency of the observed attribute would tend to a fixed limit if the observations were indefinitely continued" (*op. cit.*, 15); but his official position is that any sequence that satisfies the axioms of convergence and randomness qualifies as a collective (*op. cit.*, 28f). The interpretation hesitantly offered here is a form of the frequency theory that is in a similar way based essentially on what Popper calls virtual sequences rather than actual sequences, but it is a markedly different interpretation from either of its precursors.

The leading idea of the new theory is that the infinite sequences to be taken as collectives should be not progressions (sequences of the same shape as the natural numbers in their natural order), but dense sequences of the order type of the rational numbers (though perhaps continuous collectives may be admissible too). In the case of dicing, you are asked to put aside the image of a die thrown once a second for all eternity, and replace it with the image, more unrealistic still, of a denumerable number of throws squeezed densely into a finite interval. In experience, of course, you will meet only finite sequences. But, as before, it is required that finite sequences yield—somehow—(negative) information about the hypothetical dense sequences from which they are drawn. Let us restrict the discussion (as is customary) to binary sequences, whose two basic attributes are 0 and 1. We suppose that the positions in a dense sequence are indexed by the rational numbers in the open interval $(-1, 1)$, exactly as those in a progression are indexed by the positive integers. How are we to define the probability of the attribute 0 in the interval $(-1, 1)$? The direct analogy of the limit postulated by von Mises is obtained by calculating relative frequencies in ever larger finite subsets, and then postulating that these relative frequencies converge. The analogy of insensitivity to gambling systems is that these limits should be unaffected if the interval $(-1, 1)$ is replaced by any dense disinterestedly chosen subinterval. The analogy of Popper's axiom, on the other hand, is that the finite subsets themselves be representative of the distribution of the whole interval. The sequence of 0s and 1s, that is, should look (stochastically) similar on any scale. There is more than a passing likeness between dense collectives defined in this manner (and von Mises's collectives for that matter) and fractals (Mandelbrot 1982) All these sets of points are everywhere self-similar.

So far, there is a reasonable parallel between dense collectives and progressions. In the matter of single-case probabilities, however, the new theory can do strikingly better. Let us single out an interior point of $(-1, 1)$, which may just as well be 0, though any other would do (for in the index set it is only the ordering of the points that matters; the metric is irrelevant). How are we to define the probability of 0 at the point 0? The answer is amazingly simple, almost obvious. We define it as the limiting value (if there is a limit) of the probability of 0 in shorter and shorter dense subintervals of $(-1, 1)$ containing 0. The two limiting processes can surely be combined into one, so that, if we think of the index points as points in time, the single-case probability of an attribute at a time is the instantaneous value of its relative frequency at that time. This definition applies alike to conditional and absolute probabilities.

So much for the formalities. It cannot be too strongly urged that there are many far from trivial mathematical problems here, problems that I do not pretend to have solved. Foremost is that of showing that dense collectives exist; only when that is established will any substance have been given to the fantasy here indulged in. It does not seem that the problems are simply routine reformulations of those that were posed, and have been solved, for von Mises's theory.

Even if the above proposal can be made technically watertight, the question must be faced whether it makes good intuitive sense. The idea of throwing a die an unbounded number of times in succession is unacceptably unrealistic to some writers (Jeffrey 1977), and a dense series of throws may seem decisively worse. The subjunctive becomes imperative. Yet although dense collectives are certainly imaginary, they are locally confined; they do not, like von Mises's collectives, disappear into the remote future. It is thanks to this, I maintain, that the present theory solves the problem of single-case probabilities. For now the appropriate reference class for any such probability is determined wholly by temporal factors. Let a die be thrown at a particular instant in a prescribed (but not completely prescribed) way. Then the probability of the outcome *six-up* is defined by the frequency of *six-up* that would have emerged in throws, made in the same prescribed way, arbitrarily close in time to the given instant. The world not being limitlessly supple, approximately the same conditions, but not exactly the same conditions, govern these variant throws. The relative frequency may have a well-defined limit strictly between 0 and 1, even if the physics of the die-throwing is fully deterministic.

What about cases, such as survival throughout the next year, or radioactive decay, in which the event under scrutiny is not a response to some external triggering action (such as the throwing of the die)? Care is needed here. The probability now of the survival for the next year of a named man may, because the man persists through small intervals around the present, look destined to take the value either 0 or 1, depending only on whether the man dies or not. But this is to forget that in any frequency theory, the present one included, probabilities are defined only relative to a collective, a set of similarly generated events or of similar objects. The collective may only be virtual, or potential, but it must be announced. What distinguishes the present theory from von Mises's is that it uses the relation of contemporaneity to mark out the reference class best suited for the calculation of single-case probabilities. Hence an ensemble of co-existing insurants, or radioactive atoms, is necessary, but also sufficient, if we are to talk intelligibly of the single-case probability of survival or of decay.

There is a metaphysical picture lying behind this interpretation of probability, a picture of rapidly—even perhaps discontinuously—fluctuating background conditions. Doubtless this is an idealization; and ultimately, I suppose, it cannot be accepted as more than that (just as we cannot accept that a coastline is really a fractal). At all levels the present proposal poses as many problems as it solves—a sign, though not more than a sign, of a theory that is worth cherishing (Popper 1990, 26). It is admittedly not quite fair to ask him for solutions to these problems, but it is to Karl Popper, who in the last 60 years has made the most momentous contributions to the interpretation of probability, that I offer these fragmentary ideas as a 90th birthday present, with love and with all good wishes for many happy repetitions.

10

Truth, Truthlikeness, Approximate Truth

10.1. The Problem of Verisimilitude

The principal topic of this chapter is the impact on falsificationism of the failure of Popper's early attempt (1963, 233f) to provide a satisfactory, if very simple, theory of verisimilitude or approximation to truth; and, of course, whether later theories are any more successful. Since 1974, the year in which the difficulties in Popper's theory of verisimilitude were publicly unveiled (Tichý 1974; Miller 1974), a great deal of specialist attention has been devoted to verisimilitude, and to related problems associated with the comparison of logical and empirical contents, the possibilities of scientific progress, and the pertinence of the miracle argument for scientific realism (see 2.1.i above). It is not my intention here to survey these intricate developments. The interested reader may consult two works of Niiniluoto: his 1987 on the technical aspects, and his 1984 on the more philosophical aspects (see also the useful survey by Brink [1989], and my review [1990] of Niiniluoto 1987 and other works). My concern is only to summarize what I think falsificationism has to learn from this sorry affair, and to say briefly where I now stand. It will be clear that I do not regard the situation as a very congenial one. I shall postpone to Chapter 11 a discussion of what, to my mind, remains the most serious difficulty in the area.

Before 1974 few authors who discussed Popper's work found it necessary to mention his theory of verisimilitude. For example, Magee makes no mention of it in his *Popper,* though there is a brief discussion of truth content (1973, 42). Matters have since changed; and now few discussions of critical rationalism forbear from mentioning the problem, even if discussion is limited to a summary of Popper's own theory (as, for example, in Bouveresse 1978, 77–79), perhaps accompanied by a not-always-very-valid proof of its defects (for example, Ackermann 1976, 87–92; Newton-Smith 1981, 52–59). A 1981 study of Popper's philosophy ended with a list "of several controversies concerning 'critical rationalism' that, despite misunderstandings on all sides, cannot be regarded either as empty or as closed", the last of them being "the debate, which is essentially internal to Popper's philosophy, on the possibility of giving a precise logical meaning to the absolutely crucial idea of 'better approximation to the truth' (verisimilitude)" (Boyer 1981, 95f). Yet in this excellent account of Popper's epistemology and methodology there is no other reference to this "absolutely crucial idea". It looks almost as though verisimilitude was not seen to be of any importance for falsificationism until it was thought to pose a major embarrassment to it. There could hardly be a nicer illustration of the falsificationist doctrine that it is not the successes that a theory has that count, but its failures.

My own position is that a satisfactory theory of verisimilitude is not a crucial component of falsificationism, though such a theory is badly needed and would surely be an adornment to falsificationism. Here I tend to agree with Ackermann *loc. cit.* and with Watkins 1978, 365f; 1984, 279–88, though I am less confident than they are that the heart of the problem lies in the understanding of content rather than verisimilitude; content and verisimilitude will, I imagine, go hand in hand. The fact is that the absence of an adequate objective theory of verisimilitude hardly contradicts falsificationism, and it does not detract at all from its claim to have solved the problem of induction; what it would show is that the role of pure reason in both thought and action is more restricted than we had supposed (and, I do not mind admitting, at one time hoped). But just because of this it seems to me to be of the greatest importance that investigations concerning verisimilitude should be prosecuted with both urgency and vigour. I cannot agree with those critical rationalists for whom approximation to truth constitutes a very important and valuable idea but is not allowed to pose any problem. Albert, for example, writes (1978, 219):

> The viability of the idea of approximation to truth, which has long been effective in scientific thought, does not seem to be seriously threatened by the failure of certain attempts to provide formal definitions. The situation is similar to that of the idea of truth, the acceptability of which is not impaired by the fact that certain definitions may be controversial.

The point, of course, is that (as was true of truth before 1931) the matter is of the greatest seriousness if all theories of approximation to truth turn out to be controversial; or worse, demonstrably inadequate. I am inclined to judge that this is at present indeed the position. The matter is therefore still serious (as Popper 1979, Appendix 2, 371–74 acknowledges; but for an opinion closer to Albert's, see his 1983, xxxv–xxxvii). For perhaps there are no objective differences between hypotheses that intuitively are judged to approximate the truth to quite different degrees. We must not assume in advance that there are such differences to be found; still less should we continue to assume it uncritically when every attempt to reveal the differences has ended in failure. No doubt it was possible (and still is possible) to continue to use naive set theory after the discovery of the paradoxes; but the development of axiomatic set theory played an important part too, and helped to rescue naive set theory as a first approximation. Think of the attempts, in the latter half of the nineteenth century, to find a mechanical model of the luminiferous ether. I doubt that they should be dismissed as futile on the grounds that technical models cannot by themselves fully underwrite intuitive ideas.

The importance of the debate here has, I believe, been missed by many because of the misleading way it has been represented as an issue solely of definition (or even formal definition). Radnitzky, for example, shrugs off the whole controversy thus (1980, 204; see also Radnitzky and Andersson 1978b, 17):

> One cannot do without primitive (undefined) concepts. Which concept one wants to use as primitive is often a matter of taste, e.g., whether one chooses . . . 'more accurate representation than . . . ' or '(logical) probability'. In my opinion 'more accurate representation' (*Zutreffendere Darstellung*) is intuitively at least as clear as the concept of logical probability, especially since the former is a completely indispensable ingredient of every functioning language . . . . I must confess that for this reason the difficulties with the formal definition of 'verisimilitude' do not unsettle me greatly. Rather, the discussion evokes an impression of déjà vu: it is entirely analogous to the discussion of the formal definition of 'empirical significance'—which didn't lead to anything.

This seems to me rather to miss the point—though it may be a fair enough comment on the most well-known definitions of verisimilitude, such as those of Tichý 1976; 1978, Niiniluoto 1987, Tuomela 1979, Oddie 1986, and Schurz and Weingartner 1987, which bristle with technicalities yet do not enlighten us much about what is supposed to distinguish the more truthlike hypotheses from the less truthlike. There is, I repeat, a serious problem here, the problem of whether our intuitive judgements of truthlikeness or verisimilitude are in fact judgements of anything objective at all. It certainly has not been shown yet that this problem is insoluble. Popper's theory (though perhaps, as Zahar [1983, 167] says, "a very close formal rendering of our intuitive conception of truth-likeness") will not do, I agree: and the syntactic theories just mentioned—in addition to being unbearably complex—all suffer from the well-known disability of language dependence and for this reason seem to me to be unacceptably arbitrary. (I shall say a little more about language dependence in Note 10.4.d). But not all theories set out to give a fully explicit logical definition of verisimilitude. I agree with Radnitzky that this is unnecessary. Yet I cannot agree that because an idea is intuitively undemanding, and taken as undefined, we should therefore not bother ourselves to investigate it. Still waters run deep. In any case, the results of the last 20 years must make anyone wonder whether the intuitive idea of verisimilitude is such a clear idea after all.

If the logical problem can be solved, even faintly, there is then the further problem of whether, and on what account, we should rationally prefer hypotheses of greater verisimilitude; from the mere fact that some hypotheses are more truthlike than others it hardly follows that we should prefer the more truthlike. After that, there is the problem of whether the methods of science are well adapted, or adapted at all, to the search for increased verisimilitude. The later problems depend, of course, on a positive answer to the first. Nevertheless, I shall try to say something about each of them first, before proceeding to survey the state of the question (which is neither trivial nor pedantic) of whether there is anything in the world that corresponds, say, to our judgement that Newton's theory is a closer approximation to the whole truth than are the combined forces of Kepler's theory and Galileo's. I shall not be able to say very much about this problem, but I do hope to be able to say enough to suggest that there is no room for complacency here, but no room either for unshakable gloom. A minimal objective theory of truthlikeness does indeed seem to be possible. Whether an objective theory of scientific progress can be built on its back I do not know.

10.2. Verisimilitude as an Aim

The simplest picture of the growth of science portrays it as a succession of conjectures aimed at encapsulating the whole truth, or anyway some substantial and interesting part of the truth. (Note that this is meant to be an oversimplified picture, and intentionally disregards the undisputed fact that there is no pre-ordained language or library of languages in which all science can be formulated.) These conjectures are retained as long as they survive severe testing, but eliminated as soon as they are discovered to be false. No falsified conjecture can be regarded as a candidate for the truth. On this view any discovery that an explanatory hypothesis is false is ranked as a scientific achievement, and contributes towards the growth of science; for by falsifying the hypothesis we undoubtedly learn something about the world. On the other hand, only the invention of a true hypothesis will contribute much towards the progress of science. I stress that it is not the discovery that a hypothesis is true that is needed here—thank goodness—, nor even its survival of rigorous testing, but its invention. The discovery of true test statements, admittedly, provides some truth; but the empirical content of test statements is so small that very little progress is to be made in this way. In the simplest picture, that is, scientific progress is extremely difficult, some may say virtually impossible, to achieve. Certainly if physical science has progressed since Aristotle it is not easy to understand that progress solely in the above terms. This point has nothing to do with the complaint of Stove (1982, 3, cited at the beginning of Chapter 3 above), that according to falsificationism there has been no increase in knowledge of the traditional justificationist kind. The question here is whether science has progressed.

An adequate theory of verisimilitude would allow us to replace this picture of science with a subtler one, in which science may not only grow but also progress, despite the fact that most of the hypotheses ever invented are false. In this refined picture false hypotheses are distinguishable in principle, though only uncertainly in practice, by their verisimilitude, or their distance from the truth. With luck, at every stage there will be a theory *H*—which may have been refuted—that is judged to be the most truthlike of all those in the area of interest. If there is such a theory *H*, it is retained; not of course with any pretence to truth, if it has been refuted, but simply as a guide to where the truth may be conjectured to lie. Progress on this view is obtained by inventing a theory *K* that is closer to the truth than *H* is. Again I stress that it is not the discovery of *K*'s greater truthlikeness that is important, nor even any judgement we may make to this effect, but the invention of *K*. If some *K* is judged to be a better approximation to the truth than *H* is, it will supplant *H*. In principle, therefore, it may be possible to approach the truth through a succession of false theories.

The question must arise, however, of whether we would want to do this, even if we were able to. Popper has suggested that "the search for verisimilitude is a clearer and a more realistic aim than the search for truth" (1972/1979, 57), implying that truthlike hypotheses may be easier to obtain than true ones are, though perhaps no easier to recognize; and Koertge has gone as far as to suggest that "very great progress could be made by revising Popper's theory of methodology so as to make the aim of science be theories of ever-increasing verisimilitude, instead of truth *simpliciter*" (1978, 276). But we must not be misled here by what is a very natural, but unfortunately not correct, picture of how the verisimilitude of our hypotheses might increase—I mean the idea that we might systematically eliminate errors one by one, thus gradually approaching total freedom from error; that is, truth. What the criticisms of Popper's original theory of verisimilitude show is that any false theory is going to be packed with error: indeed, as established by Keuth 1976 and Vetter 1977 independently, a false theory has exactly as many false consequences as it has true consequences. (A summary account of these results is provided in Note 10.4.a.) The only way indeed in which error can be reduced (without a clear reduction of informative content) is by the substitution of a true hypothesis for a false one. And it is just this very restricted picture of the progress of science that we are attempting to transcend. It is accordingly not too clear why, if increase in verisimilitude is not accompanied by any reduction in error, we should prefer the more truthlike of two theories in our struggle to discover the truth.

Most writers on the subject have seen this point clearly enough. What must be important in the assessment of false hypotheses is not simply the amount of error they commit but also the seriousness of the errors committed. We seem to be driven to recognize the need not only for some measure of the verisimilitude of a hypothesis (that is, the extent to which it gets everything right), but also for a measure of its degree of truth; that is, the extent to which it gets right what it is attempting to get right. Such a measure is not easy to provide. Indeed, it is not easy to say, in the case of most false hypotheses, what it was that they were trying to say truly. Only by neglecting the unintended consequences of a scientific hypothesis—that part of its content that was not dictated in the course of the solution of the problems the hypothesis was introduced to solve—can we hope to specify the true hypothesis to which a false hypothesis is an approximation. Even in a case apparently as straightforward as Kepler's laws there is no easy way of saying what the true theory in the domain is. (See Chapter 11 for a further discussion of this point.) Certainly it is not a mere catalogue of the orbital positions of each of the planets; that would be to suggest that Kepler (along with every other theoretical scientist who has ever lived) was totally wasting his time. And, as I shall show in Chapter 11, even at the level of numerical predictions it is no routine matter to say to what degree one hypothesis approximates another. The degree of truth of a hypothesis, or the measure of the seriousness of the errors it makes, or the idea of approximate truth, seems to me to be an idea that, in strong contrast to its verisimilitude, is not even intuitively very clear.

If we are to prefer theoretically those hypotheses that in some sense make errors of less seriousness than those of their rivals, and if this preference is to be an empirical one, then it seems as if the consequences of such hypotheses will have to be similarly related. To put the matter succinctly, if slightly loosely: approximately true hypotheses must have approximately true consequences; degree of truth must not diminish in valid inference. For it is only through the consequences of a hypothesis that we are able to effect any empirical judgement. Even more does this apply in the practical domain. For we surely wish our actions to be, if not wholly appropriate to what is going to happen, at least approximately so; if not completely successful, at least approximately successful. It seems therefore that approximate truth, like truth itself, ought to be transmitted from the premises to the conclusion of a valid argument. As stated, of course, this principle is pretty vague. But it does place some constraints on theories of approximate truth (degree of truth) and distance between hypotheses, and indeed may be sufficient already to rule some of them out (Weston 1992, Section 2). I return to it, not very successfully, in Note 10.4.b below. The principle of transmission of approximate truth is not obviously correct, it should be said (Laudan 1981, 30f; 1984, 228f). Indeed, all that we have learnt about chaos and stability in recent years suggests that in general the requirement is simply false: hypotheses that, as far as we can judge, are very close together can easily have consequences that are very far apart. A clear instance is provided by the logistic function described in 8.2; another example, not involving any chaotic element, is the binomial function mentioned in 8.3. Of course it may be questioned whether two hypotheses must be accounted similar if they differ only, and by a minute amount, in the value of a single numerical quantity; nonetheless, examples such as these may make us wonder even more why approximately true hypotheses are thought to be so valuable.

When Popper first introduced and discussed the problems of verisimilitude he wrote (1963, 235):

> Let me first say that I do not suggest that the explicit introduction of the idea of verisimilitude will lead to any changes in the theory of method. On the contrary, I think that my theory of testability or corroboration by empirical tests is the proper methodological counterpart to this new metalogical idea. The only improvement is one of clarification.

To this I would say that the most rigorous falsificationist methodology, which is precisely the method that is appropriate to the search for truth, must surely be supplemented a little, if not modified, if our aim is to invent hypotheses of ever greater verisimilitude. For the extent, either in quantity or quality, to which a hypothesis errs, is something, as we have stressed, that must be open to empirical investigation. This need not at all mean departing from the injunctions of falsificationism to propose testable hypotheses and to test them severely. But it does mean that these methods must be applied in more detail than they are on the simplest falsificationist model. Not only the hypotheses at issue, but also the statements comparing their verisimilitude and their degrees of truth, will be conjectures that are urgently in need of the appropriate attempts at refutation.

10.3.   An Alternative Approach

Taking the aim of science to be the invention of theories of advanced verisimilitude is, I have suggested, an attractive proposal, but a snareful one (even if the technical problems are all solved). Is there an alternative? In this section I want to start again from scratch, look afresh at the problem of verisimilitude with untroubled eyes, and see whether, despite the mass of difficulties exposed in the last ten years or so, we cannot deal with the problem in a really simple way, as Popper originally hoped. Strangely enough, I think that perhaps we can. A simple logical model of the progress of science that preserves a good deal of its objectivity does not seem after all to be quite impossible. The model that I shall suggest is certainly no more sophisticated than Popper's first idea, and it is intuitively almost as compelling. Whether the history of science fits the model even roughly is of course another matter.

The new proposal has three components, which I state, then discuss and defend. The first component, which is—for all its innocent appearance—the crucial one, was suggested to me by Larry Briskman.

a. Truth, rather than truthlikeness or verisimilitude, is the aim of science. The way we go after the whole truth is to try to approximate it with true theories. But merely approximating the whole truth is not part of the aim of science.

b. Theory *K* is more truthlike than theory *H* is if it may be obtained from *H* EITHER by adding true propositions OR by removing false propositions without removing true propositions.

c. If we make a selection of which facts to count as central we can treat conflicting theories as empirical approximations to each other. We may then say that if *K* corrects *H* (as it may), then it is an advance over *H* provided it is more truthlike than some empirical approximation to *H*. The selectivity needed for the judgment of approximation does not infect the logical character of judgments of truthlikeness.

Let me try to defend briefly these three doctrines, which may be somewhat unexpected.

a. If *H* is a theory, Cn(*H*) is the set of its logical consequences, the consequence class or content of *H*. We assume throughout that logically equivalent propositions are identified. Cn(*H*) can be split into two parts, the truth content Ct(*H*), which is the class of all the true consequences of *H*, and the falsity content Cf(*H*), the (sometimes empty) class of its false consequences. It is easily shown that Ct(*H*) is also a theory; but normally Cf(*H*) is not. Popper's original proposal was just this: when *K* and *H* are different theories, then *K* is more truthlike than *H* is if everything that *H* gets right *K* also gets right, while everything that *K* gets wrong *H* also gets wrong; in other words, if Ct(*K*) includes Ct(*H*) and Cf(*H*) includes Cf(*K*). What Tichý and I showed about this theory of verisimilitude may be summed up as follows:

i. If *K* is true and implies *H* then *K* is more truthlike than *H* is.
ii. If *K* is false then Ct(*K*) is more truthlike than *K* is.
iii. If *K* is more truthlike than *J* is, and *J* is more truthlike than *H* is, then *K* is more truthlike than *H* is.
iv. These are all the ways in which one theory can be more truthlike than another is.

In particular, no false theory *K* can be more truthlike than some other theory *H* is. It is clear that the proposal does not give us any idea how false theories can differ in their truthlikeness. And I have no intention of trying to defend the proposal. But I should like to point out one of its neglected merits. It is quite obvious, in this theory of Popper's, why, given that science aims at truth, we should prefer from a set of competing theories that one, if there is one, that is the most truthlike. For the most truthlike theory either adds new truths to, or eliminates errors from, its rivals. So we may approach the truth by whittling away errors. Unfortunately, as we have just noted, it turns out that this can be achieved only in one whittle, instead of piecemeal.

Whatever their other merits (and demerits), most of the more recent attempts to define verisimilitude (listed above) fail to have this attractive feature. In the previous section I have emphasized that getting closer to some goal is not necessarily a useful proceeding if our real wish is simply to achieve the goal. (In Chapter 1 too I tried to explain why getting closer to certainty plays no part in the search for truth; for highly probable theories do not automatically have the one desirable property of theories that are certain, namely the property of being true.) For example, an American citizen who was not born in the United States may proceed through the various ranks of dogcatcher, city mayor, state assemblyman, state senator, state governor, congressman, U. S. senator, approaching all the time the goal of U. S. President. But constitutionally he is unable to hold this last position, so that the approach to it is quite pointless if that is his real aim. Although truth is a very clear aim, truthlikeness is by no means as clear, and cannot replace truth as our fundamental aim unless it has independent advantages of its own.

In short, we need an idea of approximation to the whole truth that is not merely ordinal. Better approximations to the truth must in some clear and useful way contain more truths. For whatever else we are doing in science, we do attempt to identify truths.

b. It turns out that there is a very simple theory that meets this requirement. At a heuristic level it is markedly similar to Popper's, and it must be conceded that it is not a lot more impressive at first sight. What is surprising is that the main results, such as they are, can be obtained by intuitively distinct routes. I think that we might be able to do something with it.

The content of a theory H, and also its truth and falsity content, were identified above with syntactic objects, classes of propositions. I propose now that we define the content of H in the standard alternative way, as the class $CN(H)$ of worlds (more correctly, models) in which H fails. Let ω be the world we are in, the single model of the set T of all truths. Clearly ω belongs to $CN(H)$ if H is false, because H fails in ω; but not if

H is true. The truth content of *H*, denoted by CT(*H*) will be the class of worlds in which *H* fails; that is, every world in CN(*H*) except possibly ω; that is, CT(*H*) = CN(*H*)\\{ω}. The falsity content CF(*H*) should be what remains; in other words, CF(*H*) is empty if *H* is true (which is what we expect), and equals {ω} if *H* is false. All false theories have the same falsity content {ω}. If we now say (directly mimicking Popper's theory) that when *K* and *H* are different theories, *K* is more truthlike than *H* is provided CT(*K*) includes CT(*H*) and CF(*H*) includes CF(*K*), then we very easily obtain the following results:

   I. If *K* is true and implies *H* then *K* is more truthlike than *H* is.

   II. If *K* is false then CT(*K*) is more truthlike than *K* is.

  II+. If *H* is false and *K* implies *H* then *K* is more truthlike than *H* is.

  III. If *K* is more truthlike than *J* is, and *J* is more truthlike than *H* is, then *K* is more truthlike than *H* is.

  IV. These are all the ways in which one theory can be more truthlike than another is.

What this amounts to is, it must be conceded, rather little: *K* is closer to the truth than *H* is if it has more truth content—except when it has also an inferior truth value. In note b. of the next section I shall sketch two intuitively dissimilar metrical theories that also yield these general results **I**, **II**, **II+**, **III**, and **IV**.

Clauses **I**, **II**, **III**, **IV** are just like **i.**, **ii.**, **iii.**, **iv.** in Popper's theory. Only **II+**, which says that a false theory is more truthlike than any of its false consequences, is new. This has the effect of turning a very weak theory of truthlikeness into one that seems unnaturally strong. For progress through a succession of false theories has suddenly been converted from a hopeless task, an impossible task, into a sinecure: given a false theory *H*, all apparently that we need to do to get closer to the whole truth is to conjoin any new proposition *g*, even a false one, to it. I have to admit that I have previously shown sympathy with this line of objection (Miller 1974, 167; 1979, 422f), and have even encouraged it. But here I want to recant, to maintain that the objection is incorrect. This is especially so when we consider the central idea of the present theory, stated above in a.

Note first that moving from a false *H* to a logically stronger *K* can always be represented as the result of the adjunction of some new truth to *H*. And it is supposed to be one of the features of the present theory that science is fundamentally in the business of identifying new truths. For if g is true, and not a consequence of the false theory H, then H-&-g is plainly the result of conjoining to H a new true proposition. But if g is false, and if f is a false consequence of H, then the new theory H-&-g can be written also in the form H-&-$(f$-iff-$g)$; and this again is the result of conjoining to H a new true proposition. But why the criticism—that progress is too easy—fails is that, as stressed in component a. of this proposal, it is not part of the aim of science to produce theories that are more truthlike than their predecessors are: what science aims to do is to discover new truths. So if we are really interested in truth we shall scorn any attempt to get closer to it simply by conjoining truths (which are always useful) or falsehoods (which often are not) to a theory already known to be false. Any new truths we can discover we shall endeavour to hold on to. But we shall want rapidly to eliminate propositions or theories that are known to be false.

This retort provokes a further problem. If our interest is in the truth, and not in truthlikeness, we shall not want to move on from a false theory to one that is more truthlike (because it would be no less false; it would share all the errors of its predecessors). In particular we require a new theory to preserve what was successful in the old one and to identify more significant truths. Normally, then, a new theory will not be more truthlike than the theory it supersedes. If we cannot accommodate some of these ideas within our account of the progress of science, then we shall not have accounted too well for the progress of science.

c. It is now well known that no two conflicting theories, unless complete, answer the same questions, or even (whatever the initial conditions might be) predict values for the same sets of numerical quantities (Miller 1975, Section II). So it is not possible for K to preserve all the good parts of H and amend all its errors, even if we restrict our attention to experimental or even numerical predictions. This is one of the cases where there really cannot be gains without accompanying losses. Yet the fact that it took logicians, not physicists, to notice this, and the development of physics itself, where predictions are usually cast in numerical form and selected ones checked by measurement, suggest that the empirical content lost from a superseded theory is often so unimportant as to be quite expendable. Matters are not so straightforward if we go above the empirical level to the theoretical; here it is often clear what has been lost, but it may be unclear whether it has been satisfactorily replaced by something of equivalent stature. The familiar problem remains, in short, of explaining the conditions under which a theory K sufficiently preserves, and amends, the content of a theory H.

I do not pretend that there is any entirely logical solution to this problem. In fact, I think that we have gone as far as pure logic will take us. The logical theory of truthlikeness just presented will be shown in Note *a.* of the next section to be a metrical theory in disguise; that is, it spells out those comparisons of truthlikeness that hold for all of a very wide class of (not necessarily numerical) distance functions ∂ defined on the class of propositions. We may therefore be able to make some progress if we superimpose on what we have so far a more determinate measure of distance $\partial(H, K)$ between theories, the idea being that $\partial(H, K)$ will be small when H and K give similar predictions across a wide range, but not the whole range, of phenomena. The details of the metric ∂ will in most cases be a matter of scientific judgment, rather than a matter of logic, and will take into account the relative importance of the various predictions made by H and K. I do not, however, rule out a substantial semantic basis to some features of ∂. Our judgments of when two theories are close together will depend on other theories, and on our interests, to be sure; but they will depend also on the way in which words are used, and much else. None of this need stop such judgments from being to a considerable extent objective, and even testable. But although the value of $\partial(H, K)$ may depend on semantic features, I should not expect it to depend on the syntactic structure of the theories H and K; $\partial(H, K)$ will be the same however H and K are formulated. This is, for all its vagueness, an algebraic theory of distance, and accordingly it is not susceptible to the charge of language dependence. On this see also Note 10.4.d.

Granted a metric ∂, we may regard K as a satisfactory replacement for H if it is close (under ∂) to some theory J that is more truthlike (in the logical sense) than H is. If H is false, J will be false too, but K need not be. Indeed, our hope will be that K is true. But even if it turns out to be false, it may be an advance on H if it corrects H in appropriate places. In other words, the progress of science is not a simple matter of approaching the truth, but one of identification of truths. It will often be the case indeed that K will be seen as an advance on H if it implies a suitable approximation to H—an idea that takes us right back to the beginning of the discussion (Popper 1961/1963, 232). Provided that ∂ satisfies the triangle inequality (which in my view any sensible measure of distance does), we shall often be able to conclude that, by the lights of the metric ∂, K is closer to the truth than H is. But it must be remembered that ∂ is not a purely logical measure of inter-theoretical distance, nor therefore of truthlikeness.

Does this non-logical aspect of ∂ matter? All realists agree that whether K is an advance on H is itself a factual matter; after all, it depends on what the whole truth T is. But most of us have, I think, held also a rather inchoate view to the effect that this is the only respect in which progress, or approach to the truth, is not logical. I call this view inchoate because the whole truth T sums up everything that is non-logical about the natural world, so to describe it as the only non-logical factor in truthlikeness is somewhat disingenuous. I am unconvinced, however, that there is any need for realists, and certainly for falsificationists, to retain this doctrine. (It is doubtless inherited from Tarski's achievement in defining formally the conditions under which a statement is true, via the doctrine of Carnap—and to some extent Keynes—that probability values are formally determinable.) Why should it not be admitted that the conditions under which K advances on H contain real factual—one hopes, testable—elements, such as the specifics of the distance function ∂? In other words, why should the values of ∂ not be provided by the scientific theories themselves? It may be thought that if K itself influences the values of ∂ then the claim that K is an advance on H will be a spurious one; for if H influenced ∂ perhaps the decision would be reversed. The possibility cannot be ruled out, of course. But this is no different from the problem of the theory-ladenness of observation, and is solved in an identical manner: assertions of scientific progress must be cast in falsifiable form if they are to be seriously entertained. (This simple falsificationist point seems to have been missed by Franklin et al. 1989.) It must be possible to refute K's claim to be a better theory than H is even on K's own terms.

Truthlikeness, in the theory here proposed, is like truth, quite objective and logical. But it is not a fundamental component of the aim of science, which is to identify truths. Whether the theory K does this better than H does cannot be objective to the same extent. At the very least it involves a methodological decision concerning which aspects of the two theories are of central importance. Should we replace H by K? This of course is an unavoidably methodological decision. Because we are interested in truth for its own sake we shall not prefer K to H if it is demonstrably not true at the time that it is proposed. But because we are not interested in truthlikeness for its own sake we may prefer K to H even if it is demonstrably not more truthlike than H is. Considerations of truthlikeness, simple-minded as they are, play an important role in guiding our decision to prefer one theory to another, because we are interested in truth. But normally they are not sufficient. Considerations of proximity between theories are needed too; we make them because we are more interested in discovering truth than we are in increasing truthlikeness.

10.4. Some Logical Notes

In this section I shall make a few technical remarks bearing on what has gone before. No attempt will be made to give a rounded picture of the state of the verisimilitude debate. Nor, I must warn the reader, will any splendid new theory of verisimilitude be unveiled here. The manifest weakness of the only theory that I can endorse will be plain enough. Yet I am not persuaded that it is so weak as to be useless. It shows how much good sense can be made of verisimilitude, even if it gives us no serious hope of application to real scientific theories. If the views put forward in the previous section are at all correct, that is the most that can be expected from a philosophical treatment of the topic.

a. Popper's Theory

There is a rather simple proof of the inadequacy of Popper's original theory of verisimilitude that gives as a corollary the later result of Keuth (1976, Section 2.3) and Vetter (1977, 372) to the effect that the truth content $Ct(H)$ and the falsity content $Cf(H)$ of a false theory H (which need not be axiomatizable) always have the same number of elements, whether that number is finite or infinite. Since this proof seems not to be well known, it is perhaps worth recording it here.

Almost all we need for the result of Keuth and Vetter is the elementary logical truism that if f is a false proposition then the proposition x and the biconditional f-iff-x always have opposite truth values. (This was the basis of my claim in Section 6 of my 1974 that the theory of propositional verisimilitude advanced by Tichý [1974] was viciously language-dependent.) So suppose that H is a false theory and f is a false consequence of H. It follows at once that if x is an element of

Ct(*H*) then *f*-iff-*x* is an element of Cf(*H*), and if *x* is an element of Cf(*H*) then *f*-iff-*x* is an element of Ct(*H*). Moreover, as may easily be checked by a truth table, *f*-iff-(*f*-iff-*x*) is logically equivalent to *x*; that is, it is *x*. In other words, the operation of taking the biconditional with *f* gives a simple association Φ of each element of Ct(*H*) with a unique element of Cf(*H*), and vice versa. This suffices to show that Ct(*H*) and Cf(*H*) have the same number of elements.

Now suppose that *K* is false, that Ct(*H*) $\subseteq$ Ct(*K*), and Cf(*K*) $\subseteq$ Cf(*H*). Let *f* be some false consequence of *K*, and so of *H*. Then if the proposition *x* belongs to Ct(*K*), the proposition *f*-iff-*x* belongs to Cf (*K*), and so Cf(*H*); hence *f*-iff-(*f*-iff-*x*), which is the same as *x*, belongs to Ct(*H*). We may conclude that Ct(*H*) = Ct(*K*). In exactly the same way, Cf(*K*) = Cf(*H*). Thus *H* and *K* are identical, and *K* cannot be more truthlike than *H* is.

b. Distance Functions

A problem raised, and immediately put to one side, in 10.2 was the problem of generalizing from truth to degrees of truth or approximate truth (as I shall for convenience call it) the truism that in a valid argument true premises lead necessarily to true conclusions. More briefly, though less exactly: do approximately true theories have consequences that are also at least approximately true? More exactly, though less briefly: do the consequences of an approximately true theory approximate the truth at least as well as the theory itself does? I must stress that the response that I am going to make to these questions, though trite enough, does not simply elucidate some trite principle of the transmission of approximate truth from premises to conclusion. Indeed it may not even be clear that it seriously comes to grips with the problem at all. An alternative approach using many-valued logic is due to Scott 1974; 1976 (see also Katz 1982), and a comparison of this with the present approach is urgently needed.

We must start by being clear what we can, and what we cannot, expect. It would be too much to expect, in particular, that the consequences of the less seriously erroneous of two theories should in every case be less seriously erroneous than the consequences of the other; that is, that if *K* is a better approximation to the truth than *H* is then every consequence of *K* should be a better approximation to the truth than every consequence of *H* is. This assuredly cannot hold in general, since (by the result of Keuth and Vetter mentioned above) half of H's consequences will be true, and thus less seriously in error than half of K's consequences, namely those that are false. The best that we can hope for is that each consequence of K is a better approximation than the corresponding consequence of H. But it is not at all easy to say what, amongst the consequences of theory K, might correspond to some particular consequence of theory H. We saw this above when we were trying to imagine what true theory Kepler's laws might be supposed to correspond to. Although a proof-theoretical elucidation of this idea seems to be in general unattainable, there can be little doubt that, at the level of predictions, there is something attractive about the idea of matching consequences that are obtained by parallel lines of proof: the way a prediction is obtained from a theory is usually independent of the initial condition with which the theory is supplemented. As we shall see, however, the attractiveness of this idea is rather superficial.

For simplicity of exposition (and also, it must be admitted, for other reasons) I shall limit the discussion here to the finite case; that is, to languages in which all theories are finitely axiomatizable. Even the class T of all truths will be assumed to be axiomatized by the single proposition t. This is a severe restriction, and has a bearing on what results can be obtained, but it will mean that we do not have to tread beyond the bounds of classical logic. Given that the results to be obtained are at best suggestive, it seems as if further complication would be a waste of everyone's time.

My technical proposal is in two parts. The first is that we introduce again a distance function or pseudometric operation ∂ on the class of propositions of the language in which we are interested. We postulate the usual conditions for a pseudometric:

$$\partial(h, h) = 0 \tag{1}$$

$$\partial(h, k) = \partial(k, h) \tag{2}$$

$$\partial(h, j) + \partial(j, k) \geq \partial(h, k) \tag{3}$$

([3] is the triangle inequality.) In terms of ∂, the truthlikeness of a hypothesis h increases as $\partial(h, t)$ decreases, whilst its degree of truth is the degree to which it approximates some unspecified truth; unspecified, because (I suspect) in general unspecifiable. The second part of the current proposal is that we initially think of matching together not corresponding consequences of two propositions—since it is unclear what this correspondence can consist in—but corresponding compounds of which they are components. In particular I suggest that the distance between propositions h and k cannot be increased, and may be diminished, if we conjoin or disjoin them with arbitrarily chosen other propositions. Formulating the proposal a little technically we may write first

$$\partial(h, k) \geq \partial(h\text{-\&-}g, k\text{-\&-}g) \qquad (4)$$

$$\partial(h, k) \geq \partial(h\text{-or-}g, k\text{-or-}g) \qquad (5)$$

Inequality (4) may be thought of as a hasty generalization of the idea that two theories cannot be made more dissimilar by conjoining to each an arbitrary initial condition. As for (5), it formulates in the naivest possible way the idea that an arbitrary consequence of h should be at least as close to some corresponding consequence of k as h is to k. These two conditions may not look very illuminating on their own, but they are not quite trivial. For example, if we assume that the elements h, g, k, . . . obey classical logic, we can actually derive a similar principle for negations:

$$\partial(h, k) = \partial(\text{not-}h, \text{not-}k). \qquad (6)$$

More generally we may want to require

$$\partial(h, k) \geq \partial(X, X[h/k]), \qquad (7)$$

where X is any compound expression involving h and $X[h/k]$ is the same expression with h replaced at zero or more places by k.

More generally still, we may postulate as constraints on ∂

$$\partial(h, h) = 0 \qquad (1)$$

$$\partial(h, k) + \partial(X, g) \geq \partial(X[h/k], g), \qquad (8)$$

which are (I hope) recognizably generalizations of the standard axioms for identity. If we add also a positivity condition

$$\text{if } \partial(h, k) = 0 \text{ then } h = k \qquad (9)$$

(which is a generalization of the law of identity of indiscernibles), then from these postulates (1), (8), (9) it is possible to derive the other axioms (2) and (3) for a pseudometric operation, and each of (4)–(7) above (Miller 1984, Section 1). Amongst the consequences of (1), (8), (9) are these:

$$\text{if } t \text{ implies } k \text{ and } k \text{ implies } h, \text{ then}$$
$$\partial(k, t) < \partial(h, t). \qquad (10)$$

if t implies not-k, then

$$\partial(k\text{-or-}t, t) < \partial(k, t) \tag{11}$$

if t implies not-h and k implies h, then

$$\partial(k, t) < \partial(h, t). \tag{12}$$

Given that the truth content Ct(k) of a proposition k in a finite language can be identified with the disjunction k-or-t, these consequences (10), (11), and (12) can be regarded as metrical versions of **I**, **II**, and **II+** respectively; and **III**, of course, holds automatically. Many other equalities and inequalities will hold for some functions ∂ that satisfy (1), (8), and (9), and not for others. But (10)–(12) tell us in effect which comparisons of degrees of truthlikeness hold for all such functions ∂.

I do not pretend for a minute that this is a deep theory, or that it avoids many familiar criticisms made against theories of logical probability. Indeed, if we write *contr* for the contradictory (or logically impossible) proposition, then we may show that the function

$$P(h) = \partial(h, contr) \tag{13}$$

behaves exactly like an absolute probability measure (Miller, *op. cit.*, Section 2). (The identification of the probability of a proposition with its distance from impossibility may perhaps offer a way of interpreting probability in logical systems more general than classical logic. But this point cannot be pursued here.) The functions P and ∂ are even interdefinable (as suggested by Radnitzky in the passage quoted in 10.1 above), since we have

$$\partial(h, k) = P(h\text{-OR-}k), \tag{14}$$

where h-OR-k is the exclusive disjunction of h and k; that is, the negation of the biconditional h-iff-k. In a genuine sense then, the present theory is equivalent to the theory of (absolute) logical probability, and suffers where it suffers. My purpose in presenting it here is only to indicate that truth really can be "located somewhere in a kind of metrical or at least topological space" (Popper 1963, 232), even if the metric or the topology is a long way from being fully logically determined. It may be mentioned that Popper's original proposal (1963) also has metrical properties (Popper 1976b, 152).

Using (14) we may easily calculate the truthlikeness $\partial(h, t)$ of a proposition h explicitly, and we shall find that it depends only on the truth value of h and its probability $P(h)$:

if h is true then $\partial(h, t) = P(h) - P(t)$

if h is false then $\partial(h, t) = P(h) + P(t)$. $\tag{15}$

Since the content of a proposition can be measured by its improbability, Oddie is quite right when he says that in this theory, as in Popper's original theory, "degrees of truthlikeness of propositions with the same truth value are determined solely by considerations of content" (1987, 25). This is perhaps to be expected, given that closeness to the whole truth *t* must mean closeness to contradiction *contr* on any metrical theory in which $\partial(h, k)$ measures the similarity between what the propositions *h* and *k* assert.

Is there any way of adding any flexibility to the theory presented here? No doubt the whole truth *T* is too big a class (Popper 1976b, 155). By understanding *T* to be not the class of all true propositions, but the class of all true theoretical (lawlike) propositions—which, of course, may hold in many worlds that are distinct from our own world ω— Kuipers (1982) has shown that we are not restricted to the meagre winnings of (10)–(12) or I, II, and II+. (For a helpful introduction, see Zandvoort 1987.) I cannot go into Kuipers's ideas here, even though they are independently valuable, and they surely add some sophistication to the simple-minded ideas of this note.

c. Degrees of Truth

There is a second way of obtaining numerical judgments of relative truthlikeness corresponding to I, II, and II+ of section 3, using not metric operations but measures of deductive dependence of the sort briefly adverted to in 3.3 (and in the technical note to Chapter 3). It is not formally independent of what we have been doing in b., though there is perhaps a genuine intuitive difference.

Again attention will be restricted to the finite case, in which *T* is axiomatized by the single proposition *t*. The measure $q(h, e)$ of the deductive dependence of hypothesis *h* on evidence *e*—which is just a minor variant of a measure first introduced by Hempel and Oppenheim 1948/1965, Section 9, and later studied by Hilpinen 1970, Section IV—intuitively assesses what proportion of the content of *h* is deductively implied by *e*. Since the common content of *h* and *e* is just the proposition *h-or-e*, and the content ct(*h*) of a proposition *h* may be measured by its improbability $1 - P(h)$, it is clear that

$$q(h|e) = \text{ct}(h\text{-or-}e)/\text{ct}(h)$$
$$= P((\text{not-}h)\text{-\&-}(\text{not-}e))/P(\text{not-}h)$$
$$= P(\text{not-}e|\text{not-}h). \tag{16}$$

If we now consider the two quantities $v(h) = q(t|h)$ and $w(h) = q(h|t)$, we see that the former is a measure of how much of the truth t is implied by h, and the latter is a measure of how much of h is implied by the truth t. Very roughly, then, $v(h)$ is a measure of the truthlikeness of h, while $w(h)$ is a measure of its degree of truth. In particular, $w(h) = 1$ whenever h is true.

This identification of $v(h)$ with the truthlikeness of h is not entirely satisfactory, since it automatically assigns the same degree of truthlikeness to a false hypothesis h and to its truth content h-or-t. But provided that $P(t)$ is not zero, these two propositions will have different degrees of truth, since $w(h$-or-$t) = 1$, whilst $w(h) = 1 - P(t)/(1 - P(h))$. We may therefore be inclined to regard truthlikeness as a combination of both $v(t)$ and $w(t)$; more precisely, to say that k is more truthlike than h is if

$$\text{either } w(h) \leq w(k) \text{ and } v(h) < v(k)$$
$$\text{or } w(h) < w(k) \text{ and } v(h) \leq v(k). \tag{17}$$

The similarity between this suggestion and Popper's original theory is perhaps more formal than substantial. But there is general agreement that truthlikeness is a property that has two components. It may be checked that each of **I**, **II**, and **II+** holds when the comparative relation of truthlikeness is defined in this manner. Again, not too surprisingly, explicit values may be derived for v and w:

$$\text{if } h \text{ is true then } v(h) = (1 - P(h))/(1 - P(t))$$
$$\text{and } w(h) = 1$$
$$\text{if } h \text{ is false then } v(h) = 1 - P(h)/(1 - P(t))$$
$$\text{and } w(h) = 1 - P(t)/(1 - P(h)). \tag{18}$$

If we calculate the value of the quotient $w(h)/v(h)$ we find that it is independent of the truth value of h:

$$\frac{w(h)}{v(h)} = \frac{1 - P(t)}{1 - P(h)}.$$

In other words, the degree of truth $w(h)$ is proportional to the quotient $v(h)/ct(h)$. It comes very close, that is, to the idea that degree of truth is truthlikeness per unit of content (compare Swinburne 1973, 213). This result continues to hold even if (as suggested at the end of the previous note) we take for T not the class of all true propositions but some weaker (axiomatizable) theory.

I must emphasize once more that I do not set much store by these primitive results. The best that can be said for them is that they hold together neatly. But they cannot be said to make much clearer the idea of the degree of truth of a false proposition.

d. Language Dependence

My proposal in b. that the values of the metric ∂ will normally have to be provided in an extra-logical manner may suggest that my own position has now effectively succumbed to one of the principal failings that I have frequently exposed in others'—namely the introduction of an ineradicable element of relativity or language-dependence. This problem affects the theories of Tichý, Oddie, Niiniluoto, Tuomela, and Schurz and Weingartner, and other theories of truthlikeness too, as well as some theories of content comparison, such as that proposed by Watkins (1984, Section 5.14). As shown in the next chapter, a formally similar problem arises when we try to compare numerical theories by their relative accuracy. But the present proposal, such as it is, is not affected. Although the charge that this or that theory of truthlikeness is language-dependent has been described by Brink as "the canonical objection against the whole verisimilitude enterprise" (1989, 186), I want strongly to resist the allegation that in propounding it I have "succeeded in showing, in a variety of ways, that the notion of truthlikeness is a non-objective notion" (Barnes 1991, 310) or that I ever questioned "the worth of the whole verisimilitude enterprise" (Brink *loc. cit.*). This I have not done. I have undoubtedly used the phenomenon to question the success of many proposed solutions to the problem of verisimilitude, but that is a different matter.

The present proposal, I insist, is not prey to my own criticism. It is, as already stated, an algebraic theory in which intertranslatable propositions are identified; the degree of truthlikeness of a proposition, and comparisons of truthlikeness between propositions are therefore fully independent of what language the propositions are formulated in. A judgement of the similarity of Newtonian theory to the truth should not depend on whether it is given a Lagrangian or a Hamiltonian or some other formulation; though it may depend on whether it is regarded primarily as a theory of celestial mechanics or as one of terrestrial mechanics. Scientists frequently reformulate their theories in different languages, and think nothing of it. It is not uncommon too for a scientific theory to be adapted to a different set of phenomena, but in such cases it is natural to suppose that we are in the presence of a new theory. It may be that by allowing the non-logical character of ∂ I am introducing what Niiniluoto terms considerations of "logical pragmatics" (1987, Section 13.4). But I prefer to think of the non-logical aspects of ∂ as straightforwardly empirical, as provided by the theories under investigation, and sharply enough formulated to be properly testable. You may say that empirical theories do not normally make explicit mention of the relative importance and centrality of their various predictions. True enough. What I am saying is that they should do so if they wish to partake in claims of better or worse approximation to the truth.

The postulates (1), (8), and (9) provide what is clearly a rudimentary theory of similarity between propositions in (finitary) classical logic; the consequences (4)–(7) indicate in particular that similarity is preserved in propositional combinations. It is therefore odd that Oddie so persistently distinguishes this approach to truthlikeness from theories such as his own—which exemplify what he calls "the likeness approach" (1987, 25f). Oddie classifies theories such as the present one as instances of "the probability-content approach". This is an accurate enough description of the general results the theory delivers, though not of its provenance, which is firmly in the domain of similarity. What distinguishes the two approaches is that in the present theory it is the similarity or likeness of propositional content between two theories (one of which may be the true theory T) that is at issue, whereas Oddie is concerned with some other aspect of similarity—something like similarity of structure. In his 1987 indeed he presents his theory as a massive generalization of the picture theory of truth. Like all attempts to introduce pictorial, and other non-linguistic, elements into the representation of scientific theories, this one leaves in the dark the fundamental question of testability, of how a picture or a structure may be tested. The tangle that Oddie's theory of truthlikeness gets into in its attempt (1986, Chapter 6) to weather the accusation of language-dependence is perhaps a fair reflection of the obscurity of the idea that it is the business of scientific theories to picture the truth rather than to state it.

11

Impartial Truth

My title, I must confess from the start, is something of a misnomer; at best, a halfhearted pun. It is not my intention in this chapter seriously to address the question of the impartiality or objectivity of truth, if only because it seems to me that little more need be said on this score. That truth in general is something that is beyond human decision, or human opinion (though not therefore something that is beyond human conjecture or human discovery), will be presumed here to be a truth that is quite beyond any human desire to have it otherwise. Once Tarski had shown that the paradoxes can be eliminated, only a confusion of truth with justified truth could have prompted anyone to think that the objectivity of truth was a controversial issue. (See, for example, Popper 1963, Chapter 10, and Popper 1972; 1979, Chapters 2 and 9; and for a contrasting opinion, Goguen 1979.) Rather I wish to investigate in greater depth than in the last chapter the difficult, important, and at times depressing question of whether there can be any such thing as partial truth; whether in addition to truth and falsity themselves there are what might be called degrees of truth, or degrees of approximation to the truth; and if there are (or even if there are not), whether science has any need for such things. That is my topic—the partialness, or lack of it, of truth, not its partiality.

Although in the previous chapter I have done my best to promote the positive view that we can make some kind of sense of degrees of truth, and that we need them, in this last chapter I want to return to being troublesome. I wish to present in a new way a result that has been in the literature for several years now (Miller 1975; see also the interesting if impressionistic paper, Darmstadter 1975), according to which we cannot persuasively make much sense of the idea that one scientific hypothesis is a better approximation to the true state of affairs than is another—unless, indeed, it happens actually to be true. Or to put the matter more modestly: if a sequence of hypotheses can constitute a sequence of finer and finer approximations to the truth, then the way in which that approximation is recognized is very different from what has normally been thought. And this spells trouble for our understanding of science, trouble that will not, I think, be dispelled by the many elaborate systems of many-valued logics, logics of approximation, fuzzy logics, and the like that have been generated in recent years to take account of the undoubted fact that our scientific predictions are quite often not quite right. For my attack is in effect an attack on what most, if not all, of these systems take for granted: namely that science is the exercise in successive approximation that its recent apologists (I mention here only Kuhn 1970, 206, since he is also amongst its detractors) have always supposed it to be. I myself do not see how it can be this, though I would assuredly like to be mistaken.

Great strides, if not great advances, have been made in recent years towards the understanding of the progress of science. And we can now at least distinguish (as indeed Hilpinen 1976 did quite early on) between various ways in which we might hope our scientific work to be progressing towards the truth. In particular we may distinguish vertical and horizontal dimensions of improvement, in a manner that I shall explain. Vertical improvement consists largely in our attempts, feeble as they are, to grasp the whole truth (as it can be formulated in any given language); the improvement is an improvement at least partly in content, but mainly in truthlikeness or in verisimilitude (the terms are used synonymously). Now no one, I trust, even Tichý and Oddie at their most evangelical (Tichý 1978, Oddie 1986), takes too seriously the extraordinary claim that the problem of verisimilitude has uncontroversially been solved—that it has been shown in what way Newton's gravitational theory, for example, is more truthlike than Galileo's (if it is). But verisimilitude is not the present business. Here I am concerned more with the possibilities for horizontal improvement in a scientific hypothesis. By this I most emphatically do not mean improvements in its scope at (say) the empirical level; for an increasing wealth of predictions, like an increase in truthlikeness, is a consequence of vertical improvement—which boils down, somehow or other, to an increase in logical and empirical content. No, what I am concerned with here is the question of fit: the way in which some hypotheses, not obviously richer in content than their rivals, nevertheless appear to fit the facts better—to permeate more insistently into the crevices of the empirical material, so to speak. In short, the extent to which one hypothesis can improve in accuracy on another, or be a better approximation to the truth than is that other, is the issue that I wish to discuss again, and in a somewhat different way.

Granted, then, that questions about truthlikeness are left to one side, we may devote ourselves to the question of degrees of truth; or, as Scott 1974, 421, and, following him, Katz 1981a; 1982, call them, degrees of error or of inexactness. Scott asks plaintively (in his 1976) whether there is a way of making sense of the many values to be found in the systems of many-valued logic. And in each of the papers mentioned he proposes that perhaps such additional truth values may indeed be understood as degrees of error, at least in the one-dimensional case (1974, 421f; 1976, 66–70). Katz too has investigated this case (for example in his 1980). In this paper I am going to voice some intuitive, rather than technical, doubts whether even this one-dimensional case can be dealt with satisfactorily (see also the pertinent comments of Smiley 1976), and some very strong technical doubts concerning the many-dimensional case (looked forward to, for example, by the reviewer of Katz 1980 in *Mathematical Reviews* 82c: 03017). I cannot properly understand degrees of error in the many-dimensional case. Perhaps others can.

I am taking it for granted that we are not going to think of degrees of error as somehow lying between truth and falsity, as though there was anywhere there for them to lie. That kind of approach to the problem, in which truth and falsity figure simply as two amongst a lot, perhaps a continuum, of other values, is a decidedly unappealing one; still worse is the kind of approach in which truth and falsity disappear altogether, and the two-valued scale is not simply supplemented, but actually replaced, by a many-valued scale. Formally such systems may have their interest, but I do not believe that they are of the slightest interest to science. What we are concerned with in science, after all, is the truth about the world; with attaining it if we can, and with somehow approximating it if we do not attain it. And our method for doing this is one that searches for falsity. Any picture of science, therefore, in which falsity disintegrates as a real option, or worse, in which both truth and falsity disappear, is simply not true to the way we do science; in particular, it leaves in the deepest shadow the way in which the results of the empirical tests we perform are supposed to impinge on the hypotheses that we propose. To conclude this little diatribe, therefore, let me make clear that for me there are only two truth values: truth and falsity. Nevertheless, it may be possible to subdivide the lower of these (falsity) more or less finely—to recognize, that is, degrees of falsity, or of error, or of approximation to the truth. Whether we can actually achieve this is however a different matter.

Errors abound in the sciences, and are cheerfully admitted; and the reason seems to be that errors are not seen to be the same kind of things as mistakes. Who ever heard of a logic of mistakes? Indeed, it is just because of this, I suspect, that attempts have often been made to insinuate degrees of error between the more stately truth values of truth and falsity. Far and away the most discussed species of errors are, of course, numerical errors; and it is to these that I shall devote most of my attention. They provide, after all, in the one-dimensional case, as simple an example as you could want for the idea of a degree of error, or degree of inexactness. You measure some quantity ϕ and obtain the value 2; I measure it and obtain the value 3. In fact the true value is just 1. It seems as clear as anything not only that my measurement is more in error than yours, but that it is twice as much in error. We have not only an ordering of errors, but a way of grading or measuring them. Equally we may say that your result approximates better to the truth—in the sense of approximating better to the true value of ϕ—than does mine; indeed, that it is twice as good an approximation to the truth. Of course scale changes would lead to different evaluations of how much more in error my measurement is than yours; but this is just the sort of thing we expect from scale changes, and cannot be thought to pose a serious problem. Indeed, nothing the least contentious, or even interesting, appears to have been said so far. There is a clear and unambiguous sense, it seems, in which some evaluations of numerical quantities are more in error, and measurably more in error, than are some others.

Here already, however, I begin to see problems and difficulties. Suppose that, unbeknownst to us, we are not in fact measuring the quantity ϕ but some related (though independent) quantity ψ. It is hard to credit that this does not sometimes happen, perhaps because there is some further factor that we have not accounted for. A simple example might be a Galilean measurement of the acceleration due to gravity that in fact measures (correctly or incorrectly) only the acceleration at a particular altitude. In such a situation it certainly looks as though there is, in addition to any strictly numerical error made, also a straightforward mistake—the mistake of supposing that it is ϕ that we have measured rather than ψ. To put the matter in sentential terms, the measurement report '$\phi = 3$' may be an incorrect report not only because 3 is the wrong value but because ϕ is the wrong quantity. It may perhaps be felt that there is no difficulty here: given that ϕ actually equals 1 the report that $\phi = 3$ is much more in error than the report that $\phi = 2$. That is fair enough. But the point that I am trying to make is that it may not be clear which truth, or true sentence, it is that our measurement report is attempting to approximate. As an answer to the question 'What is the true value of ϕ?' doubtless '$\phi = 2$' is better than '$\phi = 3$'. But as an answer to the question 'What is the value of ψ?' (which, in the context, is really the question that is being answered), it may not be. More generally we may say that the extent to which a sentence is in error must depend on what question it has been proposed to answer, and, more especially, on what is supposed to be the true sentence at which it is aimed (some writers have referred in this context to the target). Now in the case of measurements of single quantities the matter is usually, though, I have tried to suggest, not universally, straightforward. But in most other cases, especially most of those of concern to theoretical science, straightforward it usually is not.

Examples abound of scientific hypotheses and theories that are admitted to be in error, and in some clear sense merely numerically in error, for which it would not be easy to say what the true target hypothesis was. Kepler's three planetary laws provide a striking instance. They are acknowledged to be incorrect. Yet what is the truth for which they are striving? It is surely not just a catalogue of all the planetary positions, a completion as it were of Brahe's data. For such a catalogue leaves out of account, apparently, many numerical interrelationships that Kepler's laws (rightly or wrongly) succeed in bringing to light. Quite apart, therefore, from the complexity of the errors that Kepler's laws make, it seems to me to be very far from straightforward what the extent of those errors might be thought to be. In other words, it is very unclear that there is such a simple unadorned quantity as the degree of error, or degree of approximation to the truth, of Kepler's theory. Allowing ourselves some massively generous helping of idealization we may perhaps say that for any particular truth, that is, true sentence, Kepler's theory approximates that truth in some degree. To show that such things can be sensibly said is part of the burden of 10.4 above. But it does not follow from this that there is some degree to which Kepler's theory approximates the truth. Of course there is a degree to which it approximates the whole truth; that is what we call its truthlikeness or verisimilitude. But here, as I have already stressed, we are not concerned with a theory's shortcomings in content; only with its shortcomings in accuracy and in fit.

To summarize this, my first main point. Although the truth or falsity of a hypothesis is something that is quite independent of what questions that hypothesis was proposed as an answer to (provided we eliminate context-dependent elements, indexicals, and the like), this cannot be claimed of the degree of truth, or degree of error of the hypothesis. Many writers have, I am afraid, incautiously generalized from the case of evaluations of single quantities, where the matter might be presumed to be uncontentious enough, to the general case. For example, Kitcher (1993, 122): "There is no need for any very complex notion of verisimilitude to do justice to cases like this. . . . Instead of trying to achieve a linguistic ordering of statements, we let the world do the work of ordering for us. . . . Once this is done, the artificial problems that have been at the focus of much logically ingenious work on verisimilitude can be bypassed." Only if we do generalize from the one-dimensional case, of course, is there any point in having a real logic of error; if we were concerned only with numerical evaluations one at a time we could presumably restrict ourselves to equational logic, or something like it, and the apparatus of sentential and predicate calculus could be dispensed with. But once we admit that the degree of error of a hypothesis is a relative matter—that is, is relative to what is regarded as the true sentence being approximated—we shall allow too that the degree of error of a sentential compound of two hypotheses may well be no function at all of the separate degrees of error; or if it is some function, that function is likely to be an intricate one. I mention this point not because I want to follow it up here, but only to emphasize that the very simple criticism so far made may have dramatic consequences for all those many attempts to formulate logics of error based in some manner on the original many-valued logics of 70 years ago.

Happily, I myself do not believe that any such logics are necessary, and am convinced that questions concerning the approximation of one hypothesis by another can at least be made sense of within the classical tradition. One way in which this might be done is summarized in Note 10.4.b above; and in Note 10.4.c I made also a weak proposal concerning approximate truth. But although we can perhaps make sense of approximate truth in the abstract, its concrete realization is a different matter. I wish to use the remainder of this chapter to strike deeply at the presumption that every false hypothesis is true to some degree by questioning not just the absoluteness and objectivity of this supposed degree of truth but its existence altogether, even in the numerical case. I am going to present a rather simple result (first sketched in my 1975), that except in the one-dimensional case the degree of error, or degree of exactness, or degree of accuracy, of a scientific hypothesis is not invariant under trivial mathematical reformulations. More strongly, that even an ordering by relative accuracy is not so invariant. If we assess the accuracy of a prediction of the value of a quantity by the extent of its separation from the true value (as we are surely inclined to) then we shall find that in one formulation one hypothesis is uniformly more accurate and therefore presumably less in error, than another; whilst in another formulation, intertranslatable with the first, the reverse obtains. This result is exceedingly general, and will be shown to apply in almost all cases of hypotheses that seek to predict the values of more than one quantity. If degrees of error exist for such hypotheses, therefore, they do not simply vary directly with the accuracy that the hypotheses achieve.

Elementary numerical illustrations are easier to cope with than any algebraic manipulations, so I shall start with an elegant example devised by Popper (1979, 372–74). Let H and K be two rival hypotheses engaged in predicting the values of the quantities ϕ and ψ—and, we may suppose, nothing else. Suppose that their predictions, along with the values given by the true theory T, are as stated in the accompanying table:

	ϕ	ψ
H	0.150	1.225
K	0.100	1.000
T	0	1.000

It then appears that with regard to both ϕ and ψ the hypothesis K is superior in accuracy to the hypothesis H; and therefore, one assumes, is overall the less erroneous hypothesis. But now define in terms of ϕ and ψ two new quantities η and ξ as follows:

$$\eta = \psi - 2\phi$$
$$\xi = 2\psi - 3\phi$$

In terms of these new quantities the performance of H and K may be simply calculated, and stated thus:

	η	ξ
H	0.925	2.000
K	0.800	1.700
T	1.000	2.000

With regard to these quantities it is readily apparent that H is the more accurate, that is, less erroneous, hypothesis. So the relative degrees to which the hypotheses H and K are in error depends, it seems, not just on what they say but on how they say it. And there is not, on the face of it, much to choose here between the ways that they do say it. It is not as though η and ξ are an obviously contrived and unphysical pair of quantities, compared with which ϕ and ψ are refreshingly direct. Or, rather, if this is the case I do not know why it should be: for it turns out that ϕ and ψ may be redefined from η and ξ in just the way that η and ξ were defined from ϕ and ψ:

$$\phi = \xi - 2\eta$$
$$\psi = 2\xi - 3\eta.$$

Nothing could be more deliciously symmetrical. In particular cases, I am prepared to admit, there may be substantial physical reasons for preferring one of these pairs of quantities to the other. But we all know that there are scores of physical quantities—lengths, distances, velocities, masses, and the like—linear combinations of which are just as meaningful quantities in their own right. A very nice example is given by Mormann (1988, 516): in classical mechanics the Hamiltonian and the Lagrangian of a freely falling particle in a gravitational field are linearly interdefinable with the square of its momentum and its distance above the ground. These are all quantities that are as physically meaningful as you could want. Thus we can anticipate that there will be many instances where the degree of error of some simple empirical hypothesis is not straightforwardly to be determined from the approximate correctness of the predictions it makes on some selected set of quantities.

Reversals of ordering by accuracy can indeed be obtained even in the one-dimensional case if we are prepared to allow discontinuous transformations (Good 1975, 205). But this cannot be thought to be anything like as interesting, since some topological restraints must be insisted on if our reformulated hypotheses are to be reformulations at all. What is surprising is that there is so much topological variance in the midst of such a concordat on continuity. Notice, incidentally, that the actual numerical values recorded above are of scant importance: what has been shown is that although K's predictions of the values of ϕ and ψ lie between those of H and of T, in the case of the quantities η and ξ it is H's predictions that lie between those of K and of T. The result is unaffected by any change of scale for some or all of the quantities. It would not be surprising if there were more than one numerical measure of the seriousness of the errors that H and K commit: errors can of course be amplified. I am accordingly unable to see the example as just one more example of the (no doubt false) thesis that "higher dimensional physical magnitude spaces are metrically amorphous" (Mormann, *op. cit.*, 513). Still less is my argument based "on the method of measuring accuracy of an equation by taking the absolute difference between the left and right sides . . . [so that the] equations $a = b$ and $100a = 100b$ have different degrees of inaccuracy" (which is how Weston 1992, 68, elects to report the matter). What the argument points to is something much more general, to a result that is fully order-theoretic and depends on no specific numerical evaluations at all.

I now propose to show, briefly and I hope without excessive fuss, how such a generalization may be achieved. A proof of the general result was given originally in my 1975, 174–81, but there it was lopsided in character, and inelegant around the edges. The present demonstration is much to be preferred, for its symmetry as well as for its conciseness. We shall suppose, as before, that ϕ and ψ are numerical quantities, perhaps depending on independent variables $s, t, \ldots$ (which we shall systematically suppress), and that H and K are hypotheses predicting the values of these quantities. That is to say, there are numerals $\phi_H, \phi_K, \psi_H, \psi_K$ such that H implies both

$$\phi = \phi_H \text{ and } \psi = \psi_H,$$

while K implies both

$$\phi = \phi_K \text{ and } \psi = \psi_K.$$

As above we shall use T as a name for the true theory, and ϕ_T and ψ_T for the true values of ϕ and ψ, so that T implies both

$$\phi = \phi_T \text{ and } \psi = \psi_T.$$

We shall construct two simple linear combinations of ϕ and ψ in terms of which the hypotheses H and K behave in topsy-turvy ways; and for this purpose we will need further numerals λ, μ, ν. (It will be possible for λ, μ, ν, too to depend on the independent variables $s, t, \ldots$.) Let us then write down our assumption, which is that with respect to the quantities ϕ and ψ the hypothesis K is uniformly more accurate than is the hypothesis H:

$$\text{either } \phi_H < \phi_K < \phi_T \text{ or } \phi_T < \phi_K < \phi_H$$

and (1)

$$\text{either } \psi_H < \psi_K < \psi_T \text{ or } \psi_T < \psi_K < \psi_H.$$

The case where some of these inequalities might become equalities (as in Popper's illustration above) can be accommodated, but it is necessary that enough proper inequalities remain to ensure two things: first that H and K are distinct hypotheses, and second that K and T are distinct (that is to say, K is not true as regards ϕ and ψ). As we shall see, we shall need to require also that ϕ and ψ are genuinely distinct quantities, so that we do not collapse back to the one-dimensional case.

Now the condition that a linear combination $\eta = \phi + \lambda\psi$ should reverse H and K in accuracy; that is,

$$\text{either } \phi_K + \lambda\psi_K < \phi_H + \lambda\psi_H < \phi_T + \lambda\psi_T$$
$$\text{or } \quad \phi_T + \lambda\psi_T < \phi_H + \lambda\psi_H < \phi_K + \lambda\psi_K \tag{2}$$

may be written in the form

$$\text{if } \quad \phi_K + \lambda\psi_K \lessgtr \phi_H + \lambda\psi_H$$
$$\text{then } \phi_H + \lambda\psi_H \lessgtr \phi_T + \lambda\psi_T. \tag{3}$$

So whether or not $\psi_H < \psi_K$ or $\psi_K < \psi_H$ it follows that

$$\text{if } \frac{\phi_K - \phi_H}{\psi_H - \psi_K} \le \lambda \text{ then } \lambda \le \frac{\phi_H - \phi_T}{\psi_T - \psi_H}. \tag{4}$$

We may therefore choose for λ some weighted average of these two fractions, say

$$\lambda = \mu . \frac{\phi_K - \phi_H}{\psi_H - \psi_K} + (1-\mu) . \frac{\phi_H - \phi_T}{\psi_T - \psi_H}, \tag{5}$$

where $0 < \mu < 1$. Provided the two fractions being weighted are not actually equal,

$$\frac{\phi_K - \phi_H}{\psi_H - \psi_K} \ne \frac{\phi_H - \phi_T}{\psi_T - \psi_H}, \tag{6}$$

in which case we would have equality almost everywhere, we can proceed to define a new quantity η by

$$\eta = \phi + \lambda . \psi \tag{7}$$

and with respect to η the hypothesis H will be more accurate than the hypothesis K is. In a similar way we can write

$$\nu = \mu . \frac{\psi_K - \psi_H}{\phi_H - \phi_K} + (1-\mu) \cdot \frac{\psi_H - \psi_T}{\phi_T - \phi_H}, \tag{8}$$

and define a new quantity ξ by

$$\xi = \psi + \nu . \phi.$$

In terms of η, ξ we may then redefine ϕ, ψ by noting that

$$\phi = \eta - \lambda\psi = \eta - \lambda\xi + \lambda\nu\phi \tag{9}$$

so that,

$$\phi = \frac{\lambda\xi - \eta}{\lambda\nu - 1} \text{ and } \psi = \frac{\nu\eta - \xi}{\lambda\nu - 1}. \tag{10}$$

A highly symmetrical set of definitions is obtained with

$$\eta = \frac{\lambda\psi + \phi}{\sqrt{(\lambda\nu - 1)}} \text{ and } \xi = \frac{\nu\phi + \psi}{\sqrt{(\lambda\nu - 1)}}$$

$$\phi = \frac{\lambda\xi - \eta}{\sqrt{(\lambda\nu - 1)}} \text{ and } \psi = \frac{\nu\eta - \xi}{\sqrt{(\lambda\nu - 1)}} \tag{11}$$

where $\lambda\nu$ is the product of the terms on the right-hand sides of equations (5) and (8). It can be shown that $\lambda\nu$ exceeds ≥ 1; and that it equals 1, thus inciting a discontinuity somewhere, if and only if the condition (6) fails; that is to say,

$$\frac{\phi_K - \phi_H}{\phi_H - \phi_T} = \frac{\psi_K - \psi_H}{\psi_H - \psi_T}. \tag{12}$$

Equation (12) will hold, for example, when ϕ is some linear function of ψ, and in other cases too where their independence as quantities will seriously be in question.

Given that (12) does hold, our construction will be unsuccessful. Nevertheless it is usually easy enough to provide an alternative construction that will perform the same trick. We may, for example, if ϕ and ψ are non-negative, replace them by ϕ^2 and ψ^2. Our assumption that K is more accurate than H will not be affected by this substitution. Now the analogous condition to (12), namely

$$\frac{\phi^2_K - \phi^2_H}{\phi^2_H - \phi^2_T} = \frac{\psi^2_K - \psi^2_H}{\psi^2_H - \psi^2_T}, \tag{13}$$

will hold together with (12) only if

$$\frac{\phi_T}{\psi_T} = \frac{\phi_H}{\psi_H} = \frac{\phi_K}{\psi_K}. \tag{14}$$

that is, only if ϕ is a numerical multiple of ψ. In such a circumstance it cannot be said that within the context of the hypotheses H and K the quantities ϕ and ψ are seriously distinct.

To illustrate the construction let us take a simple example where each of ϕ and ψ is a function of an independent variable t. The predictions issued by H, K, and T are as follows:

	$\phi(t)$	$\psi(t)$
H	t	$5mt$
K	$t + a$	$2mt$
T	$t + 2a$	mt

where a and m are constants, or, anyway, are not themselves functions of t. (It is assumed that the constant a is of the same dimension as t.) With respect to the quantities $\phi(t)$ and $\psi(t)$ the hypothesis K is manifestly more accurate than the hypothesis H is. Yet trivial calculations show that $\lambda = 5a/12mt$, that $\nu = 5mt/2a$, and thus that $\lambda\nu - 1 = \frac{1}{24}$. The new quantities take a form that is only deceptively complicated:

$$\eta(t) = \frac{5a\sqrt{24}}{12mt}\,\psi(t) + \sqrt{24}\cdot\phi(t)$$

$$\tag{15}$$

$$\xi(t) = \frac{5mt\sqrt{24}}{2a}\phi(t) + \sqrt{24}\cdot\psi(t)$$

and the predictions made for them (after division throughout by $\sqrt{24}$) are as follows:

	$\eta(t)/\sqrt{24}$	$\xi(t)/\sqrt{24}$
H	$t + 25a/12$	$5mt^2/2a + 5mt$
K	$t + 22a/12$	$5mt^2/2a + 9mt/2$
T	$t + 29a/12$	$5mt^2/2a + 6mt$

As was to be expected, H is the more accurate hypothesis for each of η and ξ throughout the whole range of t.

Our construction, though mathematically elementary, is obviously of considerable generality. Yet Horton 1978 is quite right to insist that it is not as radical as it appears at first sight. For there are precious few examples in the history of science where one competing hypothesis turns out to be uniformly more accurate than another, even on a single quantity; and in most real cases, therefore, other considerations (such as functional form) have to be brought to bear if we are to express any empirical preference between the competing hypotheses. (For a related point, see Franklin 1988.) And the case that I have considered can be seen as just a special instance of this much more general phenomenon.

Though for the quantity ϕ the hypothesis K may be uniformly more accurate than H over the range of t, when the weighting factor μ is allowed to vary as well a very different picture emerges. I cannot go into Horton's discussion in any detail here; I shall merely emphasize, as he does, that though the result may be less radical it is not therefore any less serious. For just as there is no such relation between hypotheses of one's being uniformly the more accurate, so there is no relation of one's being preponderantly the more accurate. From a formal point of view false hypotheses that predict values for the same set of quantities are on a par. (There is another difficulty here that I pass over; see my 1975, 166f.) If one such false hypothesis is preferred to another, then presumably one family of quantities is preferred to another. In my 1975, 181–86, I made faces at such preferences, accusing them of essentialism, anyway in the general case. That was no doubt an excessive reaction, as Mormann *op. cit.* justly observes. But what enthusiasts of the idea that our scientific hypotheses do approximately explain the facts (such as Girill 1978), or that they approach the truth through successive approximation or observational nesting (such as Newton-Smith 1981 206f), must surely do is to come forward with some way of coping with the difficulty. As Mormann makes plain (*op. cit.,* 518), one cannot simply disregard what is, after all, a plain and straightforward result of something like high-school mathematics.

Nothing that I have said here proves, of course, that there cannot be degrees of error, or degrees of inexactness, or (looking at it from the other end) degrees of truth. All that can be said is that, if they exist, they depend on factors of which we have at present only a limited understanding. From a methodological point of view their presence (and in some sense openness to empirical investigation) is of the most considerable importance. This point was made incessantly in the previous chapter. Despite the insipid discussion of Haack 1980, especially pages 15–17, there seems to me to be the strongest possible methodological demand for our being able to make sense in some way of the degree to which some hypothesis, in answer to some problem, approaches what might be thought to be the true answer to that problem. I am no less sorry than I was 20 years ago that I still seem unable to do much at all to satisfy that demand.

BIBLIOGRAPHY

ACHINSTEIN, Peter. 1968. Review of POPPER 1963. *British Journal for the Philosophy of Science* 19: 159–168.

ACKERMANN, Robert. 1976. *The Philosophy of Karl Popper*. Amherst: University of Massachusetts Press.

AGASSI, Joseph. 1975. *Science in Flux*. Dordrecht: Reidel.

AGASSI, Joseph, and JARVIE, Ian Charles, eds. 1987. *Rationality: The Critical View*. Dordrecht: Nijhoff.

ALBERT, Hans. 1978. Science and the Search for Truth. In RADNITZKY and ANDERSSON 1978a.

ASQUITH, Peter D., and HACKING, Ian, eds. 1978. *PSA 1978*, Vol. 1. East Lansing: Philosophy of Science Association.

AUSTIN, J. L. 1962. *How to Do Things with Words*. Oxford: Oxford University Press.

AYER, Alfred J. 1956. *The Problem of Knowledge*. Harmondsworth: Penguin Books.

———. 1957. The Conception of Probability as a Logical Relation. In KÖRNER 1957.

———. 1972. *The Central Questions of Philosophy*. London: Weidenfeld and Nicholson.

———. 1973. *Probability and Evidence*. London: Macmillan.

———. 1982. *Philosophy in the Twentieth Century*. London: Weidenfeld and Nicholson.

BACON, Francis. 1620. *Novum Organum*. London: Billium. References are to Book and numbered aphorism.

BALDWIN, J. M., ed. 1901. *Dictionary of Philosophy and Psychology*, Vol. 2. London: Macmillan.

BARNES, Eric. 1990. The Language Dependence of Accuracy. *Synthese* 84: 59–95.

——. 1991. Beyond Verisimilitude: A Linguistically Invariant Basis for Scientific Progress. *Synthese* 88: 309–339.

BARNES, Jonathan. 1990. *The Toils of Scepticism*. Cambridge: Cambridge University Press.

BARTLEY, William Warren, III. 1962. *The Retreat to Commitment*. London: Chatto and Windus.

——. 1964. Rationality versus the Theory of Rationality. In BUNGE 1964.

——. 1971. *Morality and Religion*. London: Macmillan.

——. 1973. *Wittgenstein*. London: Quartet. 2d ed. 1985 La Salle: Open Court.

——, ed. 1977. *Lewis Carroll's Symbolic Logic*. Hassocks: Harvester. 2d ed. 1986 New York: Potter.

——. 1978. *Werner Erhard: The Transformation of a Man*. New York: Potter.

——. 1981. Eine Lösung des Goodman-Paradoxons. In RADNITZKY and ANDERSSON 1981.

——. 1984. *The Retreat to Commitment*. 2d ed., rev. and enl. La Salle: Open Court.

——. 1987. A Refutation of the Alleged Refutation of Comprehensively Critical Rationalism. In RADNITZKY and BARTLEY 1987.

——. 1990. *Unfathomed Knowledge, Unmeasured Wealth: On Universities and the Wealth of Nations*. La Salle: Open Court.

BERKSON, William. 1979. Skeptical Rationalism. *Inquiry* 22: 281–320. Reprinted in abridged form in AGASSI and JARVIE 1987.

——. 1990a. In Defense of Good Reasons. *Philosophy of the Social Sciences* 20: 84–91.

——. 1990b. Methodology is Pragmatic: A Response to Miller. *Philosophy of the Social Sciences* 20: 95–98.

BERTOCCI, Peter A., ed. 1974. *Mid-Twentieth Century Philosophy: Personal Statements*. New York: Humanities Press.

BINNS, Peter. 1978. The Supposed Asymmetry between Falsification and Verification. *Dialectica* 32: 29–40.

BLACK, Max. 1983. *The Prevalence of Humbug, and Other Essays*. Ithaca: Cornell University Press.

BOHNEN, Alfred, and MUSGRAVE, Alan, eds. 1991. *Wege der Vernunft: Festschrift zum siebzigsten Geburtstag von Hans Albert*. Tübingen: Mohr.

BOON, Louis. 1979. Repeated Tests and Repeated Testing: How to Corroborate Low Level Hypotheses. *Zeitschrift für allgemeine Wissenschaftstheorie* 10: 1–10.

BOUVERESSE, Renée. 1978. *Karl Popper ou Le rationalisme Critique*. Paris: Vrin.

BOYER, Alain. 1981. La tyrannie de la certitude. *Esprit* (May 1981): 69–97.

BRINK, Chris. 1989. Verisimilitude: Views and Reviews. *History and Philosophy of Logic* 10: 181–201.

BRISKMAN, Larry. 1977. Historicist Relativism and Bootstrap Rationality. *The Monist* 60: 509–539.

————. 1983. *Problems and their Progress*. Ph.D. dissertation, University of Edinburgh.

BUNGE, Mario, ed. 1964. *The Critical Approach to Science and Philosophy*. New York: Free Press.

————, ed. 1967. *Quantum Theory and Reality*. Berlin: Springer.

BUTTS, Robert E., and HINTIKKA, Jaakko, eds. 1977a. *Foundational Problems in the Special Sciences*. Dordrecht: Reidel.

BUTTS, Robert E., and· HINTIKKA, Jaakko, eds. 1977b. *Basic Problems in Methodology and Linguistics*. Dordrecht: Reidel.

CALDWELL, Bruce J. 1991. Clarifying Popper. *Journal of Economic Literature* XXIX: 1–33.

CARNAP, Rudolf. 1950. *Logical Foundations of Probability*. Chicago: University of Chicago Press.

CARNAP, Rudolf, and JEFFREY, Richard C., eds. 1980. *Studies in Inductive Logic and Probability*, Vol. 2. Berkeley: University of California Press.

CARROLL, Lewis. 1895. What the Tortoise Said to Achilles. *Mind* 4:289f. Reprinted in COPI and GOULD 1982.

CLAASSEN, E. M., ed. 1967. *Les Fondements Philosophiques des Systèmes Economiques*. Paris: Bibliothèque Economique et Politique.

COHEN, L. Jonathan. 1977. *The Probable and the Provable*. Oxford: Clarendon.

————. 1978a. Is Popper More Relevant than Bacon for Scientists? *Times Higher Education Supplement* (July 14, 1978): 11.

————. 1978b. What Scientists Cannot Learn from Popper. *Times Higher Education Supplement* (September 1, 1978): 11.

————. 1980a. What Has Science to Do with Truth? *Synthese* 45:489–510.

————. 1980b. Review of NATHAN 1980. *Times Literary Supplement* (March 14, 1980): 302.

COHEN, L. Jonathan, and HESSE, Mary, eds. 1980. *Applications of Inductive Logic*. Oxford: Clarendon.

COHEN, Robert S., FEYERABEND, Paul K., and WARTOFSKY, Marx W., eds. 1976. *Essays in Memory of Imre Lakatos*. Dordrecht: Reidel.

COPI, Irving M., and GOULD, James A., eds. 1982. *Readings on Logic*. 2d ed. New York: Macmillan.

da COSTA, Newton C. A., and DORIA, Francisco Antonio. 1991. Undecidability and Incompleteness in Classical Mechanics. *International Journal of Theoretical Physics* 30: 1041–1073.

da COSTA, Newton C. A., and DORIA, Francisco Antonio. 1992. On the Incompleteness of Axiomatized Models for the Empirical Sciences. *Philosophica* 50: 73–100.

da COSTA, Newton C. A., and FRENCH, Steven. 1989. Pragmatic Truth and the Logic of Induction. *British Journal for the Philosophy of Science* 40:333–356.

D'AGOSTINO, Fred, and JARVIE, I. C., eds. 1989. *Freedom and Rationality: Essays in Honor of John Watkins*. Dordrecht: Kluwer.

DARMSTADTER, Howard. 1975. Better Theories. *Philosophy of Science* 42:20–27.

DELAHAYE, Jean-Paul. 1993. Randomness, Unpredictability, and Absence of Order: the Identification by the Theory of Recursivity of the Mathematical Notion of Random Sequence. In DUBUCS 1993a.

DIACONIS, Persi, and ZABELL, Sandy L. 1982. Updating Subjective Probability. *Journal of the American Statistical Association* 77: 822–830.

DORLING, Jon. 1979. Bayesian Personalism, the Methodology of Scientific Research Programmes, and Duhem's Problem. *Studies in the History and Philosophy of Science* 10: 177–187.

————. 1981. Bayesian Personalism, Falsificationism, and the Problem of Induction. *Aristotelian Society Supplementary Volume* 55: 109–125.

DREBEN, Burton, and HEIJENOORT, Jean van. 1986. Introductory Note to *1929, 1930* and *1930a*. In GÖDEL 1986.

DUBUCS, Jacques-Paul, ed. 1993a. *Philosophy of Probability*. Dordrecht: Kluwer.

DUBUCS, Jacques-Paul. 1993b. Inductive Logic Revisited. In DUBUCS 1993a.

DUHEM, Pierre. 1906. *La Théorie Physique: son Object, sa Structure*. 2d ed. Paris: Rivière, 1914.

————. 1954. *The Aim and Structure of Physical Theory*. English translation of DUHEM 1906 by P. P. WIENER. Princeton: Princeton University Press.

EARMAN, John. 1992. *Bayes or Bust? A Critical Examination of Bayesian Confirmation Theory.* Cambridge, Ma: MIT Press.

EDWARDS, W., LINDMAN, H., and SAVAGE, L. J. 1963. Bayesian Statistical Inference for Psychological Research. *Psychological Review* 70: 193–242.

EELLS, Ellery. 1988. On the Alleged Impossibility of Inductive Probability. *British Journal for the Philosophy of Science* 39: 111–16.

ETCHEMENDY, John. 1990. *The Concept of Logic Consequence.* Cambridge, Ma: Harvard University Press.

FEIGL, Herbert. 1974. No Pot of Message. In BERTOCCI 1974. Reprinted in FEIGL 1981.

———. 1981. *Inquiries and Provocations: Selected Writings 1929–1974.* Dordrecht: Reidel.

FELLER, William. 1968. *An Introduction to Probability Theory and Its Applications.* 3d ed. New York: Wiley.

FETZER, James H. 1981. *Scientific Knowledge: Causation, Explanation, and Corroboration.* Dordrecht: Reidel.

FEYERABEND, Paul K. 1968. A Note on Two 'Problems' of Induction. *British Journal for the Philosophy of Science* 19: 251–53.

———. 1978. In Defence of Aristotle: Comments on the Condition of Content Increase. In RADNITZKY and ANDERSSON 1978a.

———. 1981. More Clothes from the Emperor's Bargain Basement. *British Journal for the Philosophy of Science* 32: 57–71.

de FINETTI, Bruno. 1937. La Prévision: ses Lois Logiques, ses Sources Subjectives. *Annales de l'Institut Henri Poincaré* 7: 1–68.

———. 1964. Foresight: Its Logical Laws, Its Subjective Sources. English translation of de FINETTI 1937 by Henry E. KYBURG, Jr. In KYBURG and SMOKLER 1964.

———. 1970. *Teoria delle Probabilità.* Turin: Einaudi.

———. 1974. *Theory of Probability. Volume 1.* English translation of de FINETTI 1970 by Antonio MACHÍ and Adrian SMITH. London: Wiley.

FRANKLIN, Allan. 1988. How Nancy Cartwright Tells the Truth. *British Journal for the Philosophy of Science* 39: 527–29.

FRANKLIN, A., et al. 1989. Can a Theory-Laden Observation Test the Theory? *British Journal for the Philosophy of Science* 40: 229–231.

FREEMAN, Derek. 1983. Inductivism and the Test of Truth: A Rejoinder to Lowell D. Holmes and Others. *Canberra Anthropology* 6: 101–192.

GAIFMAN, Haim. 1985. On Inductive Support and Some Recent Tricks. *Erkenntnis* 22: 5–21.

GÄRDENFORS, Peter, and SAHLIN, Nils-Eric. 1982. Unreliable Probabilities, Risk Taking, and Decision Making. *Synthese* 53: 361–386.

GÄRDENFORS, Peter, and SAHLIN, Nils-Eric. 1983. Decision Making with Unreliable Probabilities. *British Journal of Mathematical and Statistical Psychology* 36: 240–251.

GIERE, Ronald N. 1973. Objective Single-Case Probabilities and the Foundations of Statistics. In SUPPES et al. 1973.

GIRILL, T. R. 1978. Approximative Explanation. In ASQUITH and HACKING 1978.

GLYMOUR, Clark. 1980. *Theory and Evidence.* Princeton: Princeton University Press.

GÖDEL, Kurt. 1929. Über die Vollständigkeit des Logikkalküls. Doctoral dissertation, University of Vienna. References are to the English translation, 'On the Completeness of the Calculus of Logic', *1929,* in GÖDEL 1986.

———. 1958. Über eine bisher noch nicht benützte Erweiterung des finiten Standpunktes. *Dialectica* 12: 280–87. References are to the English translation, 'On a Hitherto Unutilized Extension of the Finitary Standpoint', *1958* in GÖDEL 1990.

———. 1986. *Collected Works. Volume I, Publications 1929–1936.* New York: Oxford University Press.

———. 1990. *Collected Works. Volume II, Publications 1938–1974.* New York: Oxford University Press.

GOGUEN, J. A. 1979. Fuzzy Sets and the Social Nature of Truth. In GUPTA, RAGRADE, and YAGER 1979.

GOOD, I. J. 1967. On the Principle of Total Evidence. *British Journal for the Philosophy of Science* 17: 319–321. Reprinted as Chapter 17 of GOOD 1983.

———. 1968. Corroboration, Explanation, Evolving Probability, Simplicity and a Sharpened Razor. *British Journal for the Philosophy of Science* 19:123–143.

———. 1971. 46656 Varieties of Bayesians. *American Statistician* 25 (December 1971): 62f. Reprinted as Chapter 3 of GOOD 1983.

———. 1975. Explicativity, Corroboration, and the Relative Odds of Hypotheses. *Synthese* 30: 39–73. Reprinted as Chapter 15 of GOOD 1983.

———. 1975. Comments on David Miller. *Synthese* 30:205f.

———. 1983. *Good Thinking: The Foundations of Probability and Its Applications.* Minneapolis: University of Minnesota Press.

———. 1984. Comment on POPPER and MILLER 1983. *Nature* 310 (August 2 1984):434.

———. 1985. Probabilistic Induction is Inevitable. *Journal of Statistical Computation and Simulation* 20: 323f, C216.

———. 1987. A Reinstatement, in Response to Gillies, of Redhead's Argument in Support of Induction. *Philosophy of Science* 54: 470–72.

————. 1989/1990. A Suspicious Feature of the Popper/Miller Argument. *Journal of Statistical Computation and Simulation* 31 (1989): 62f. Reprinted in *Philosophy of Science* 57 (1990): 535f.

GOODMAN, Nelson. 1955/1965. *Fact, Fiction, and Forecast.* 2d ed. Indianapolis: Bobbs-Merrill.

GRAVES, Paul R. 1989. The Total Evidence Theorem for Probability Kinematics. *Philosophy of Science* 56: 317–324.

GRÜNBAUM, Adolf. 1976a. Is Falsifiability the Touchstone of Scientific Rationality? Popper versus Inductivism. In COHEN, FEYERABEND, and WARTOFSKY 1976.

————. 1976b. Is the Method of Bold Conjectures and Attempted Refutations *Justifiably* the Method of Science? *British Journal for the Philosophy of Science* 27: 105–136.

GUPTA, M.M., RAGRADE, R.K., and YAGER, R.R., eds. 1979. *Advances in Fuzzy Set Theory and Applications.* Amsterdam: North-Holland.

HAACK, Susan. 1979. Epistemology *with* a Knowing Subject. *Review of Metaphysics* 33: 309–335.

————. 1980. Is Truth Flat or Bumpy? In MELLOR 1980.

HACKING, Ian, ed. 1981. *Scientific Revolutions.* Oxford: Oxford University Press.

HADAMARD, Jacques. 1898. Les surfaces à courbures opposées et leurs lignes géodésiques. *Journal de Mathématiques pures et appliquées* IV (5th series): 27–73.

HAMBLIN, C. L. 1970. *Fallacies.* London: Methuen.

HARPER, William L., and KYBURG, Henry E. 1968. The Jones Case. *British Journal for the Philosophy of Science* 18: 247–251.

HARRÉ, Rom. 1986. *Varieties of Realism: A Rationale for the Natural Sciences.* Oxford: Blackwell.

HARSANYI, John C. 1985. Acceptance of Empirical Statements: A Bayesian Theory without Cognitive Utilities. *Theory and Decision* 18:1–30.

HATTIANGADI, J. N. 1987. *How is Language Possible?* La Salle: Open Court.

HAYEK, Friedrich A., ed. 1935a. *Collectivist Economic Planning: Critical Studies on the Possibilities of Socialism.* London: Routledge.

————. 1935b. The Present State of the Debate. In HAYEK 1935a.

van HEIJENOORT, Jean, ed. 1967. *From Frege to Gödel. A Source Book in Mathematical Logic.* Cambridge, Ma: Harvard University Press.

HEMPEL, Carl G. 1965. *Aspects of Scientific Explanation: Essays in the Philosophy of Science.* New York: The Free Press.

HEMPEL, Carl G., and OPPENHEIM, Paul. 1948. Studies in the Logic of

Explanation. *Philosophy of Science* 15: 135–175. Reprinted in HEMPEL 1965.

HESSE, Mary. 1974. *The Structure of Scientific Inference*. London: Macmillan.

HILBERT, David. 1905. Über die Grundlagen der Logik und der Arithmetik. In *Verhandlungen des Dritten Internationalen Mathematiker-Kongresses in Heidelberg vom 8. bis 13. August 1904*, 174–185. Leipzig: Teubner. References are to the English translation, 'On the Foundations of Logic and Arithmetic', in van HEIJENOORT 1967.

HILPINEN, Risto. 1968. *Rules of Acceptance and Inductive Logic*. Amsterdam: North-Holland.

———. 1970. On the Information Provided by Observations. In HINTIKKA and SUPPES 1970.

———. 1976. Approximate Truth and Truthlikeness. In PRZEŁECKI, SZANIAWSKI, and WÓJCICKI 1976.

HINTIKKA, Jaakko, and SUPPES, Patrick, eds. 1970. *Information and Inference*. Dordrecht: Reidel.

HINTIKKA, Jaakko, NIINILUOTO, Ilkka, and SAARINEN, Esa, eds. 1978. *Essays on Mathematical and Philosophical Logic* Dordrecht: Reidel.

HORTON, Deryck. 1978. Accuracy of Prediction: A Note on David Miller's Problem. *British Journal for the Philosophy of Science* 29: 179–183.

HOWSON, Colin. 1977. Why Once May Be Enough. *Australasian Journal of Philosophy* 55: 142–46.

———. 1989. On a Recent Objection to Popper and Miller's 'Disproof' of Probabilistic Induction. *Philosophy of Science* 56: 675–680.

———. 1990. Some Further Reflections on the Popper-Miller Disproof of Probabilistic Induction. *Australasian Journal of Philosophy* 68: 221–28.

———. 1993. Personalistic Bayesianism. In DUBUCS 1993a.

HOWSON, Colin, and FRANKLIN, Allan. 1985. A Bayesian Analysis of Excess Content and the Localisation of Support. *British Journal for the Philosophy of Science* 36: 425–431.

HOWSON, Colin, and URBACH, Peter. 1989. *Scientific Reasoning: The Bayesian Approach*. La Salle: Open Court.

HOWSON, Colin, and WORRALL, John. 1974. The Contemporary State of Philosophy of Science in Britain. *Zeitschrift für allgemeine Wissenschaftstheorie* 5: 363–374.

HUDSON, James L. 1974. Logical Subtraction. *Analysis* 35: 130–35.

HÜBNER, Kurt. 1978. Some Critical Comments on Current Popperianism on the Basis of a Theory of System Sets. In RADNITZKY and ANDERSSON 1978a.

HUME, David. 1739. *A Treatise of Human Nature. Volume I. Of the Understanding.* London: John Noon. Edition of L.A. Selby-Bigge, Oxford University Press, Oxford, 1888; 2d ed. of P.H. Nidditch, Clarendon Press, Oxford 1978.

HUMPHREYS, Paul. 1985. Why Propensities Cannot Be Probabilities. *Philosophical Review* XCIV: 557–570.

JACKSON, Frank. 1984. Petitio and the Purpose of Arguing. *Pacific Philosophical Quarterly* 75: 27–37.

JACOB, Pierre, ed. 1989. *L'âge de la science, Vol. 2: Epistémologie.* Paris: Odile Jacob.

JEFFREY, Richard C. 1965. *The Logic of Decision.* 2d ed., 1983. New York: McGraw-Hill.

———. 1975. Probability and Falsification: Critique of the Popper Program. *Synthese* 30: 95–117.

———. 1977. Mises Redux. In BUTTS and HINTIKKA 1977b.

———. 1984. Comment on POPPER and MILLER 1983. *Nature* 310 (August 2, 1984): 433.

JONES, Gary, and PERRY, Clifton. 1982. Popper, Induction and Falsification. *Erkenntnis* 18: 97–104.

KATZ, Michael. 1980. Inexact Geometry. *Notre Dame Journal of Formal Logic* 21: 521–535.

———. 1981. Two Systems of Multi-valued Logic for Science. In *Proceedings of the 11th International Symposium on Multiple-valued Logic,* 175–182.

———. 1982. Real-valued Models with Metric Equality and Uniformly Continuous Predicates. *Journal of Symbolic Logic* 47: 772–792.

KEUTH, Herbert. 1976. Verisimilitude or the Approach to the Whole Truth. *Philosophy of Science* 43: 311–336.

KEYNES, J.M. 1921. *A Treatise on Probability.* London: Macmillan.

KITCHER, Philip. 1993. *The Advancement of Science.* New York: Oxford University Press.

KLAPPHOLZ, K., and AGASSI, J. 1959. Methodological Prescriptions in Economics. *Economica* 26 (New Series): 60–74.

KLINE, Morris. 1980. *Mathematics: The Loss of Certainty.* Oxford: Oxford University Press.

KÖRNER, Stephan, ed. 1957. *Observation and Interpretation.* London: Butterworth. Reprinted 1962 by Dover, New York.

———, ed. 1976. *Philosophy of Logic.* Oxford: Blackwell.

KOERTGE, Noretta. 1978. Towards a New Theory of Scientific Inquiry. In RADNITZKY and ANDERSSON 1978a.

KOLMOGOROV, A.N. 1933. Grundbegriffe der Wahrscheinlichkeitsrechnung. *Ergebnisse der Mathematik* 2.

――――. 1950. *Foundations of the Theory of Probability.* English translation of KOLMOGOROV 1933. New York: Chelsea.

KUHN, Thomas S. 1962. *The Structure of Scientific Revolutions.* 2d ed., 1970. Chicago: University of Chicago Press.

KUIPERS, Theo A.F. 1982. Approaching Descriptive and Theoretical Truth. *Erkenntnis* 18: 343–378.

――――. 1983. Non-Inductive Explication of Two Inductive Intuitions. *British Journal for the Philosophy of Science* 34: 209–223.

――――, ed. 1987. *What Is Closer-to-the-Truth?* Amsterdam: Rodopi.

KYBURG, Henry E., Jr. 1970. *Probability and Inductive Logic.* London: Macmillan.

KYBURG, Henry E., Jr., and SMOKLER, Howard E., eds. 1964. *Studies in Subjective Probability.* New York: Wiley.

LAKATOS, Imre, ed. 1968a. *The Problem of Inductive Logic.* Amsterdam: North-Holland.

――――. 1968b. Changes in the Problem of Inductive Logic. In LAKATOS 1968a.

――――. 1970. Falsification and the Methodology of Scientific Research Programmes. In LAKATOS and MUSGRAVE 1970. Reprinted in LAKATOS 1978.

――――. 1974. Popper on Demarcation and Induction. In SCHILPP 1974. Reprinted in LAKATOS 1978.

――――. 1978. *The Methodology of Scientific Research Programmes.* Philosophical Papers, Volume 1. Edited by John WORRALL and Gregory CURRIE. Cambridge: Cambridge University Press.

LAKATOS, Imre, and MUSGRAVE, Alan, eds. 1970. *Criticism and the Growth of Knowledge.* Cambridge: Cambridge University Press.

LAUDAN, Larry. 1981. A Confutation of Convergent Realism. *Philosophy of Science* 48: 19–49. Reprinted in LEPLIN 1984.

LEHRER, Keith. 1980. Truth, Evidence, and Error: Comments on Miller. In COHEN and HESSE 1980.

LEPLIN, Jarrett, ed. 1984. *Scientific Realism.* Berkeley: University of California Press.

LEVI, Isaac. 1967a. *Gambling with Truth.* New York: Knopf.

――――. 1967b. Probability Kinematics. *British Journal for the Philosophy of Science* 18: 197–209.

———. 1969. If Only Jones Knew More! *British Journal for the Philosophy of Science* 20: 153–59.

———. 1974. On Indeterminate Probabilities. *Journal of Philosophy* 71: 391–418.

———. 1977. Epistemic Utility and the Evaluation of Experiments. *Philosophy of Science* 44: 368–386. Reprinted as Chapter 6 of LEVI 1984.

———. 1984a. *Decisions and Revisions*. Cambridge: Cambridge University Press.

———. 1984b. Comment on POPPER and MILLER 1983. *Nature* 310 (August 2, 1984): 433.

LEVINSON, Paul, ed. 1982. *In Pursuit of Truth: Essays in Honor of Karl Popper's 80th Birthday*. New York: Humanities Press.

LEVISON, Arnold. 1974. Popper, Hume, and the Traditional Problem of Induction. In SCHILPP 1974.

LEWIS, Clarence Irving. 1946. *An Analysis of Knowledge and Valuation*. La Salle: Open Court.

LEWIS, David K. 1980. A Subjectivist's Guide to Objective Chance. In CARNAP and JEFFREY 1980.

LIEBERSON, Jonathan. 1982. The Romantic Rationalist. *New York Review of Books* XXIX (December 2, 1982): 53–58.

———. 1983. Reply to MILLER 1983b. *New York Review of Books* XXX (April 28, 1983): 43f.

LINDLEY, D. V. 1965. *Introduction to Probability and Statistics*. Cambridge: Cambridge University Press.

MACKENZIE, Jim. 1985. No Logic Before Friday. *Synthese* 63: 329–341.

MACKIE, J.L. 1973. *Truth, Probability, and Paradox*. Oxford: Clarendon.

MAGEE, Bryan. 1973. *Popper*. London: Fontana.

MAHER, Patrick. 1990. Why Scientists Gather Evidence. *British Journal for the Philosophy of Science* 41: 103–119.

MANDELBROT, Benoit. 1982. *The Fractal Geometry of Nature*. San Francisco: Freeman.

MAXWELL, Grover. 1974. Corroboration without Demarcation. In SCHILPP 1974.

———. 1975. Induction and Empiricism: A Bayesian-Frequentist Alternative. In MAXWELL and ANDERSON 1975.

MAXWELL, Grover, and ANDERSON, Robert M., Jr. , eds. 1975. *Induction, Probability, and Confirmation*. Minneapolis: University of Minnesota Press.

MAXWELL, Mary Lou, and SAVAGE, C. Wade, eds. 1989. *Science, Mind, and Psychology: Essays in Honor of Grover Maxwell.* Lanham: University Press of America.

MAXWELL, Nicholas. 1993. Induction and Scientific Realism: Einstein Versus van Fraassen. Part One: How to Solve the Problem of Induction. *British Journal for the Philosophy of Science* 44: 61–79.

McGEE, Vann. 1992. Two Problems with Tarski's Theory of Consequence. *Proceedings of the Aristotelian Society* XCII (New Series): 273–292.

MEDAWAR, Peter B. 1985. *Plato's Republic.* Oxford: Oxford University Press.

MELLOR, D.H. 1971. *The Matter of Chance.* Cambridge: Cambridge University Press.

———. 1977. The Popper Phenomenon. *Philosophy* 52: 195–202.

———, ed. 1980. *Prospects for Pragmatism.* Cambridge: Cambridge University Press.

MENDELSON, Elliott. 1964. *Introduction to Mathematical Logic.* Princeton: Van Nostrand.

MILL, John Stuart. 1843. *A System of Logic, Ratiocinative and Inductive.* 8th ed., 1961. London: Longmans.

———. 1979. *The Logic of the Moral Sciences.* La Salle: Open Court. Books VI–VIII of MILL 1843.

MILLER, David. 1974. Popper's Qualitative Theory of Verisimilitude. *British Journal for the Philosophy of Science* 25: 166–177.

———. 1975. The Accuracy of Predictions. *Synthese* 30: 159–191.

———. 1978. On Distance from the Truth as a True Distance. In HINTIKKA, NIINILUOTO, and SAARINEN 1978.

———. 1980. Can Science Do without Induction? In COHEN and HESSE 1980. Replies to comments appear on 143–146, 154f.

———. 1981. Bayesian Personalism, Falsificationism, and the Problem of Induction. *Aristotelian Society Supplementary Volume* 55, 127–141.

———, ed. 1983a. *A Pocket Popper.* London: Fontana. U.S. edition 1985, *Popper Selections.* Princeton: Princeton University Press.

———. 1983b. Letter to the Editors. *New York Review of Books* XXX (April 28, 1983): 43.

———. 1984. A Geometry of Logic. In SKALA, TERMINI, and TRILLAS 1984.

———. 1990a. Some Logical Mensuration. *British Journal for the Philosophy of Science* 41: 281–290.

———. 1990b. Reply to Zwirn and Zwirn. *Cahiers du CREA,* No. 14, École Polytechnique, Paris, 149–153.

———. 1994. À quoi sert la logique? *Hermès,* 14, December 1994 (forthcoming).

MILLER, David, and POPPER, Karl. 1986. Deductive Dependence. *Actes IV Congrés Català de Lògica*. Universitat Politècnica de Catalunya & Universitat de Barcelona, 21–29.

MILNE, Peter. 1986. Can There Be a Realist Single-case Interpretation of Probability? *Erkenntnis* 25: 129–132.

von MISES, Richard. 1928. *Wahrscheinlichkeit, Statistik, und Wahrheit*. 3d ed., 1951. Vienna: Julius Springer.

————. 1957. *Probability, Statistics, and Truth*. London: Allen and Unwin, and 2d ed. 1957, New York: Macmillan. English translation of von MISES 1951.

MONK, Ray. 1990. *Ludwig Wittgenstein: The Duty of Genius*. London: Cape.

MORMANN, Thomas. 1988. Are All False Theories Equally False? A Remark on David Miller's Problem and Geometric Conventionalism. *British Journal for the Philosophy of Science* 39: 505–519.

MOTT, Peter L. 1980. Haack on Fallibilism. *Analysis* 40: 177–183.

MURA, Alberto. 1990. When Probabilistic Support Is Inductive. *Philosophy of Science* 57: 278–289.

————. 1992. *La Sfida Scettica: Saggio sul problema logico dell'induzione*. Pisa: ETS Editrice.

MUSGRAVE, Alan. 1975. Popper and 'Diminishing Returns' from Repeated Tests. *Australasian Journal of Philosophy* 53, 248–253.

————. 1988. The Ultimate Argument for Scientific Realism. In NOLA 1989.

————. 1989a. Deductivism versus Psychologism. In NOTTURNO 1989.

————. 1989b. Saving Science from Scepticism. In D'AGOSTINO and JARVIE 1989.

————. 1991. What is Critical Rationalism? In BOHNEN and MUSGRAVE 1991.

————. 1993. *Common Sense, Science, and Scepticism: A Historical Introduction to the Theory of Knowledge*. Cambridge: Cambridge University Press.

NAGEL, Ernest, SUPPES, Patrick, and TARSKI, Alfred, eds. 1961. *Logic, Methodology, and Philosophy of Science: Proceedings of the 1960 International Congress*. Stanford: Stanford University Press.

NATHAN, N.M.L. 1980. *Evidence and Assurance*. Cambridge: Cambridge University Press.

NIINILUOTO, Ilkka. 1984. *Is Science Progressive?* Dordrecht: Reidel.

————. 1987. *Truthlikeness*. Dordrecht: Reidel.

NIINILUOTO, Ilkka, and TUOMELA, Raimo. 1973. *Theoretical Concepts and Hypothetico-Inductive Inference*. Dordrecht: Reidel.

NIINILUOTO, Ilkka, and TUOMELA, Raimo, eds. 1979. *The Logic and Epistemology of Scientific Change*. Acta Philosophica Fennica XXX, Nos 2–4.

NOLA, Robert, ed. 1989. *Relativism and Realism in Science*. Dordrecht: Kluwer.

NOTTURNO, Mark, ed. 1989. *Perspectives on Psychologism*. Leiden: E. J. Brill.

ODDIE, Graham. 1986. *Likeness to Truth*. Dordrecht: Reidel.

————. 1987. The Picture Theory of Truthlikeness. In KUIPERS 1987.

O'HEAR, Anthony. 1975. Rationality of Action and Theory-Testing in Popper. *Mind* LXXXIV (New Series): 273–76.

————. 1980. *Karl Popper*. London: Routledge.

————. 1989. *An Introduction to the Philosophy of Science*. Oxford: Clarendon.

ÖPIK, U. 1957. Contribution to the discussion of AYER 1957.

PEIRCE, C.S.S. and LADD-FRANKLIN, C. 1901. *Petito Principii*. In BALDWIN 1901.

PLATO. Ed. W.K.C. Guthrie. 1956. *Plato: Protagoras and Meno*. Harmondsworth: Penguin.

POINCARÉ, Henri. 1902. *La Science et l'hypothèse*. Paris: Flammarion.

————. 1913. *Dernières Pensées*. Paris: Flammarion.

————. 1952. *Science and Hypothesis*. English translation of POINCARÉ 1903. New York: Dover.

————. 1963. *Mathematics and Science: Last Essays*. English translation of POINCARÉ 1913. New York: Dover.

POPPER, Karl R. 1933. Ein Kriterium des empirischen Charakters theoretischer Systems. *Erkenntnis* 3: 426f. English translation, 'A Criterion of the Empirical Character of Theoretical Systems', POPPER 1959a, 312–14.

————. 1934. *Logik der Forschung*. Vienna: Springer.

————. 1940. What Is Dialectic? *Mind* 49 (New Series): 403–426. Revised version, Chapter 15 of POPPER 1963.

————. 1944. The Poverty of Historicism II (Part III). *Economica* 11: 119–137.

————. 1945a. The Poverty of Historicism III (Part IV). *Economica* 12: 69–89.

————. 1945b. *The Open Society and Its Enemies*. Vol. 1: *The Spell of Plato*. Vol. 2: *The High Tide of Prophecy: Hegel, Marx, and the Aftermath*. London: Routledge.

————. 1957a. *The Poverty of Historicism*. Reprint of POPPER 1944 and 1945a. London: Routledge.

————. 1957b. The Propensity Interpretation of the Calculus of Probability, and the Quantum Theory. In KÖRNER 1957. Reprinted as selection 15, 'Propensities, Probabilities, and the Quantum Theory', in MILLER 1983a.

————. 1959a. *The Logic of Scientific Discovery*. English translation of POPPER 1934. London: Hutchinson.

————. 1959b. The Propensity Interpretation of Probability. *British Journal for the Philosophy of Science* 10: 25–42.

————. 1960. On the Sources of Knowledge and of Ignorance. *Proceedings of the British Academy* 46: 39–71. Reprinted as Chapter 1 of POPPER 1963/1989.

————. 1961. Some Comments on Truth and the Growth of Knowledge. In NAGEL, SUPPES, and TARSKI 1961.

————. 1963. *Conjectures and Refutations.* 5th rev. ed. 1989. London: Routledge.

————. 1966. *The Open Society and Its Enemies.* 5th ed. of POPPER 1945. London: Routledge.

————. 1967a. Quantum Mechanics without 'The Observer'. In BUNGE 1967. Reprinted with revisions and additions in POPPER 1982b, 35–95.

————. 1967b. La rationalité et le statut du principe de rationalité. In CLAASSEN 1967. French translation of Selection 29, 'The Rationality Principle', in MILLER 1983a.

————. 1972. *Objective Knowledge.* Oxford: Clarendon.

————. 1974a. Intellectual Autobiography. In SCHILPP 1974. Reprinted as POPPER 1976a.

————. 1974b. Replies to My Critics. In SCHILPP 1974.

————. 1976a. *Unended Quest.* London: Fontana.

————. 1976b. A Note on Verisimilitude. *British Journal for the Philosophy of Science* 27: 147–159.

————. 1979. *Objective Knowledge.* Revised ed. of POPPER 1972. Oxford: Clarendon.

————. 1981. The Present Significance of Two Arguments of Henri Poincaré. *Methodology and Science* 14: 260–264.

————. 1982a. *The Open Universe. An Argument for Indeterminism.* Vol. 2 of *The Postscript to The Logic of Scientific Discovery,* ed. W.W. BARTLEY, III. London: Hutchinson.

————. 1982b. *Quantum Theory and the Schism in Physics.* Vol. 3 of *The Postscript to The Logic of Scientific Discovery,* ed. W.W. BARTLEY, III. London: Hutchinson.

————. 1983. *Realism and the Aim of Science.* Vol. 1 of *The Postscript to The Logic of Scientific Discovery,* ed. W.W. BARTLEY, III. London: Hutchinson.

————. 1989. *Changing Our View of Causality: A World of Propensities.* Little Rock: University of Arkansas for Medical Sciences.

————. 1990. *A World of Propensities.* Bristol: Thoemmes.

POPPER, Karl R., and ECCLES, John C. 1977. *The Self and Its Brain.* Berlin: Springer International.

POPPER, Karl, and MILLER, David. 1983. A Proof of the Impossibility of Inductive Probability. *Nature* 302 (April 21, 1983): 687f.

———. 1984. Reply to LEVI 1984, JEFFREY 1984, and GOOD 1984. *Nature* 310 (August 2, 1984): 434.

———. 1987. Why Probabilistic Support Is Not Inductive. *Philosophical Transactions of the Royal Society of London* A 321, (April 30, 1987): 569–591.

POST, John F. 1987. A Gödelian Theorem for Theories of Rationality. In RADNITZKY and BARTLEY 1987.

PRIEST, Graham, ROUTLEY, Richard, and NORMAN, Jean, eds. 1989. *Paraconsistent Logic. Essays on the Inconsistent.* Munich: Philosophia.

PRZEŁECKI, Marian, SZANIAWSKI, Klemens, and WÓJCICKI, Ryszard, eds. 1976. *Formal Methods in the Methodology of Empirical Sciences.* Dordrecht: Reidel.

PUTNAM, Hilary. 1974. The 'Corroboration' of Theories. In SCHILPP 1974. Reprinted in HACKING 1981.

QUINE W.V. 1969a. *Ontological Relativity and Other Essays.* New York: Columbia University Press.

———. 1969b. Natural Kinds. In RESCHER et al. 1969. Reprinted in QUINE 1969a.

RADNITZKY, Gerard. 1980. From Justifying a Theory to Comparing Theories and Selecting Questions. *Revue Internationale de Philosophie* Nos. 131–32, 179–228.

RADNITZKY, Gerard, and ANDERSSON, Gunnar, eds. 1978a. *Progress and Rationality in Science.* Dordrecht: Reidel.

———. 1978b. Objective Criteria of Scientific Progress? Inductivism, Falsificationism, and Relativism. In RADNITZKY and ANDERSSON 1978a.

———, eds. 1981. *Voraussetzungen und Grenzen der Wissenschaft.* Tübingen: Mohr.

RADNITZKY, Gerard, and BARTLEY, W.W., III, eds. 1987. *Evolutionary Epistemology, Rationality, and the Sociology of Knowledge.* La Salle: Open Court.

RAMSEY, F. P. Undated. Manuscripts 005-20-01, 005-20-03. Archives of Scientific Philosophy in the Twentieth Century, Hillman Library, University of Pittsburgh. Reprinted in *British Journal for the Philosophy of Science* 41 (1990): 2f, with a preamble by Niels-Eric Sahlin.

RESCHER, Nicholas, et al., ed. 1969. *Essays in Honor of Carl G. Hempel: A Tribute on the Occasion of His Sixty-fifth Birthday.* Dordrecht: Reidel.

ROSENKRANTZ, Roger D. 1977. *Inference, Method, and Decision.* Dordrecht: Reidel.

———. 1983. Why Glymour *Is* a Bayesian. In EARMAN 1983.

ROTHMAN, Kenneth J., ed. 1988. *Causal Inference*. Chestnut Hill: Epidemiology Resources Inc.

RUELLE, David. 1989. *Chaotic Evolution and Strange Attractors*. Cambridge: Cambridge University Press.

RUSSELL, Bertrand. 1918. Lectures on Logical Atomism. *The Monist* 28: 495–527; 29: 32–63, 190–222, 345–380.

———. 1956. *Logic and Knowledge: Essays 1901–1950*. London: Allen and Unwin.

———. 1985. *The Philosophy of Logical Atomism*. Ed. David Pears. La Salle: Open Court.

SALMON, Wesley C. 1967. *The Foundations of Scientific Inference*. Pittsburgh: University of Pittsburgh Press.

———. 1968. The Justification of Inductive Rules of Inference. In LAKATOS 1968a.

———. 1969. Partial Entailment as a Basis for Inductive Logic. In RESCHER et al. 1969.

———. 1975. Confirmation and Relevance. In MAXWELL and ANDERSON 1975.

———. 1978. Unfinished Business: The Problem of Induction. *Philosophical Studies* 33: 1–19.

———. 1981. Rational Prediction. *British Journal for the Philosophy of Science* 32: 115–125.

SAPIRE, David. 1991. General Causation. *Synthese* 86: 321–347.

SAVAGE, C. Wade, ed. 1990. *Scientific Theories*. Minneapolis: University of Minnesota Press.

SAVAGE, L. J. 1954. *The Foundations of Statistics*. New York: Wiley. Rev. ed. 1972. New York: Dover.

SCHILPP, Paul Arthur, ed. 1974. *The Philosophy of Karl Popper*. Library of Living Philosophers, Vol. 14. La Salle: Open Court.

SCHLESINGER, George N. 1988. There's a Fascination Frantic in Philosophical Fancies. In ROTHMAN 1988.

SCHURZ, Gerard, and WEINGARTNER, Paul. 1987. Verisimilitude Defined by Relevant Consequence-elements: A New Reconstruction of Popper's Original Idea. In KUIPERS 1987.

SCOTT, Dana. 1974. Completeness and Axiomatizability in Many-valued Logic. *Proceedings of the Tarski Symposium*. Proceedings of the Symposia in Pure Mathematics XXV: 411–435. Providence: American Mathematical Society.

———. 1976. Does Many-valued Logic Have Any Use? In KÖRNER 1976.

SEIDENFELD, Teddy. 1979. Why I Am Not an Objective Bayesian: Some Reflections Prompted by Rosenkrantz. *Theory and Decision* 11: 413–440.

———. 1991. When Normal and Extensive Form Decisions Differ. Preprint, July 1991.

SEIDENFELD, Teddy, and WASSERMAN, Larry. 1991. Dilation for Sets of Probabilities. Technical Report No. 537, Department of Statistics, Carnegie Mellon University, October 1991.

SEXTUS EMPIRICUS. *Outlines of Pyrrhonism*. References are to *Sextus Empiricus* I, translation by R.G. Bury. 1961. London: Heinemann and Cambridge, Ma: Harvard University Press.

SHEARMUR, Jeremy. 1984. Review of LEVINSON 1982. *Technology and Culture* XX: 694–96.

SKALA, Heinz J., TERMINI, S., and TRILLAS, E. 1984. *Aspects of Vagueness*. Dordrecht: Reidel.

SKYRMS, Brian. 1987a. On the Principle of Total Evidence with and without Observation Sentences. In WEINGARTNER and SCHURZ 1987.

———. 1987b. Updating, Supposing, and Maxent. *Theory and Decision* 22: 225–246.

———. 1990. The Value of Knowledge. In SAVAGE 1990.

SMILEY, Timothy J. 1976. Comment on SCOTT 1976. In KÖRNER 1976.

NEWTON-SMITH, W.H. 1980. Comment on MILLER 1980. In COHEN and HESSE 1980.

———. 1981. *The Rationality of Science*. London: Routledge.

SMULLYAN, Raymond M. 1987. *Forever Undecided: A Puzzle Guide to Gödel*. New York: Knopf.

———. 1992. *Gödel's Incompleteness Theorems*. New York: Oxford University Press.

STEWART, Ian. 1989. *Does God Play Dice?* Oxford: Blackwell.

STOVE, David. 1982. *Popper and After: Four Modern Irrationalists*. Oxford: Pergamon.

STRAWSON, P. F. 1950. Truth. *Aristotelian Society Supplementary Volume* 24, 129–156. Reprinted in STRAWSON 1971.

———. 1971. *Logico-Linguistic Papers*. London: Methuen.

SUDBURY, A. W. 1973. Could There Exist a World Which Obeyed No Scientific Laws? *British Journal for the Philosophy of Science* 24: 39f.

———. 1973. Scientific Laws that are Neither Deterministic nor Probabilistic. *British Journal for the Philosophy of Science* 27: 307–315.

SUPPES, Patrick. 1987. Propensity Representations of Probability. *Erkenntnis* 26: 335–358.

SUPPES, Patrick, HENKIN, Leon, JOJA, Athanase, and MOISIL, Gr. C., eds. 1973. *Logic, Methodology, and Philosophy of Science IV*. Amsterdam: North-Holland.

SWINBURNE, Richard. 1973. *An Introduction to Confirmation Theory*. London: Methuen.

THEOCHARIS T., and PSIMOPOULOS, M. 1987. Where Science Has Gone Wrong. *Nature* 329 (August 15, 1987): 595–98.

TICHÝ, Pavel. 1974. On Popper's Definitions of Verisimilitude. *British Journal for the Philosophy of Science* 25: 155–160.

———. 1976. Verisimilitude Redefined. *British Journal for the Philosophy of Science* 27: 25–42.

———. 1978. Verisimilitude Revisited. *Synthese* 38: 175–196.

TROELSTRA, A. S. 1990. Introductory Note to *1958* and *1972*. In GÖDEL 1990.

TRUSTED, Jennifer. 1979. *The Logic of Scientific Inference*. London: Macmillan.

TUOMELA, Raimo. 1979. Scientific Change and Approximation. In NIINILUOTO and TUOMELA 1979.

URBACH, Peter. 1987. *Francis Bacon's Philosophy of Science*. La Salle: Open Court.

VETTER, Hermann. 1977. 'A New Concept of Verisimilitude. *Theory and Decision* 8: 369–75.

VILLE, J. 1939. *Étude critique de la notion de collectif*. Paris: Gauthier-Villars.

VINCENT, R. H. 1962. Popper on Qualitative Confirmation and Disconfirmation. *Australasian Journal of Philosophy* 40: 159–166.

WALTON, Douglas N. 1991. *Begging the Question: Circular Reasoning as a Tactic of Argumentation*. London: Greenwood.

WARNOCK, G. J. 1960. Review of POPPER 1959a. *Mind* 69 (New Series): 99–101.

WATKINS, J. W. N. 1958. Confirmable and Influential Metaphysics. *Mind* 67 (New Series): 344–365.

———. 1975. Metaphysics and the Advancement of Science. *British Journal for the Philosophy of Science* 26: 91–121.

———. 1978. Corroboration and the Problem of Content-Comparison. In RADNITZKY and ANDERSSON 1978a.

———. 1984. *Science and Scepticism*. Princeton: Princeton University Press.

———. 1987. Comprehensively Critical Rationalism: A Retrospect. In RADNITZKY and BARTLEY 1987.

WEINGARTNER, Paul, and SCHURZ, Gerard, eds. 1987. *Logic, Philosophy of Science, and Epistemology*. Vienna: Hölder-Pichler-Tempsky.

WESTON, Thomas. 1992. Approximate Truth and Scientific Realism. *Philosophy of Science* 59: 53–74.

WHEELER, John Archibald. 1977. Genesis and Observership. In BUTTS and HINTIKKA 1977a.

WHEWELL, William. 1847. *The Philosophy of the Inductive Sciences, Founded upon their History.* 2d ed. Vol. 2. London: Parker. Reprinted 1966 by The Johnson Reprint Corporation, New York.

WILLIAMS, Bernard. 1978. *Descartes: The Project of Pure Enquiry.* Hassocks: Harvester, and Harmondsworth: Penguin.

WORRALL, John. 1989a. Structural Realism: The Best of Both Worlds? *Dialectica* 43: 99–124.

WORRALL, John. 1989b. Why Both Popper and Watkins Fail to Solve the Problem of Induction. In D'AGOSTINO and JARVIE 1989.

ZAHAR, Elie. G. 1982. The Popper-Lakatos Controversy. *Fundamenta Scientiae* 3: 21–54. Alternative version, ZAHAR 1983.

———. 1983. The Popper-Lakatos Controversy in the Light of 'Die beiden Grundprobleme der Erkenntnistheorie'. *British Journal for the Philosophy of Science* 34: 149–171.

ZANDVOORT, Henk. 1987. Verisimilitude and Novel Facts. In KUIPERS 1987.

ZWIRN, Denis, and ZWIRN, Hervé P. 1989. L'argument de Popper et Miller contre la justification probabiliste de l'induction. In JACOB 1989.

———. 1993. Logique inductive et soutien probabiliste. *Dialogue* XXXII: 293–307.

INDEX

Index